Wireless Power Transfer for Unmanned Aircraft

This book presents the principles of wireless power transfer for unmanned aerial vehicles, from theoretical modelling to practical applications, providing a complete technical perspective and hands-on experience. It combines in-depth theoretical models, such as T-models and M-models, with practical system design, including wireless charging system construction. It presents systematic solutions to real-world challenges in UAV wireless charging, including mutual inductance disturbances and lightweight units. Providing the resources to tackle complex industry problems, this book covers the latest technological insights and advanced control methods, such as PT-symmetric wireless power transfer system control schemes and charging range extension techniques. Ideal for professional engineers, designers, and researchers, it provides the tools needed for innovation in UAV technologies and power systems. Whether you're developing new systems or optimizing existing ones, this comprehensive resource delivers the insights and techniques needed to drive progress in wireless power transfer for unmanned aircraft.

Zhen Zhang is Professor in the School of Electrical and Information Engineering at Tianjin University. He is Associate Editor for *IEEE Transactions on Industrial Electronics*, *IEEE Transactions on Industrial Informatics*, and *IEEE Industrial Electronics Magazine*. He is the recipient of the 2020 Outstanding Paper Award for *IEEE Transactions on Industrial Electronics* and the IEEE J. David Irwin Early Career Award.

Yu Gu received his BEng degree in automation from Tianjin University in 2018, and his PhD in control science and engineering from Tianjin University in 2024. Currently he serves as a postdoctoral fellow in the School of Electrical and Information Engineering at Tianjin University. He is the Young Scientist Editor of the SCI journal *Wireless Power Transfer*. His current research interests include wireless power transfer and UAV hovering wireless charging.

Qi Li received his MSc from Tianjin University and joined the School of Electrical and Information Engineering of Tianjin University in 2015. He has received the Shen Zhikang Teaching Award from Tianjin University. His research areas include embedded computing and control and intelligent robotic systems.

Wireless Power Transfer for Unmanned Aircraft

Fundamentals, Design, and Control

ZHEN ZHANG
Tianjin University

YU GU
Tianjin University

QI LI
Tianjin University

CAMBRIDGE
UNIVERSITY PRESS

Shaftesbury Road, Cambridge CB2 8EA, United Kingdom

One Liberty Plaza, 20th Floor, New York, NY 10006, USA

477 Williamstown Road, Port Melbourne, VIC 3207, Australia

314–321, 3rd Floor, Plot 3, Splendor Forum, Jasola District Centre, New Delhi – 110025, India

103 Penang Road, #05-06/07, Visioncrest Commercial, Singapore 238467

Cambridge University Press is part of Cambridge University Press & Assessment,
a department of the University of Cambridge.

We share the University's mission to contribute to society through the pursuit of
education, learning and research at the highest international levels of excellence.

www.cambridge.org
Information on this title: www.cambridge.org/9781108472432
DOI: 10.1017/9781108560405

First published 2026

Cover image: Oselote / iStock / Getty Images Plus

A catalogue record for this publication is available from the British Library

A Cataloging-in-Publication data record for this book is available from the Library of Congress

ISBN 978-1-108-47243-2 Hardback

Contents

Acknowledgements

The authors would like to acknowledge the many exceptional contributions to the content of this book from their research team – the Tianjin University Laboratory of Embedded Computing and Control (TJU-ECC) – particularly Mr. Yantian Gong, Mr. Shen Shen, Mr. Kaibo Chen, Mr. Fei Gao, and Mr. Jiahao Bao. In particular, a large portion of the supporting materials presented are the findings of the authors' research team.

The authors are deeply indebted to their tutors, colleagues, and friends worldwide for their continuous support and encouragement over the years. A note of gratitude to the editors and staff at Cambridge University Press, who were instrumental in undertaking a diligent review of the text and editing the book through the production process.

The authors would like to express their gratitude to the National Natural Science Foundation of China (Grant No. 52377104) for its financial support.

Last but not least, the authors also owe debts of gratitude to their families, who gave tremendous support during the process of writing this book.

1 Fundamentals of Wireless Power Transfer

Wireless power transfer (WPT) is a revolutionary technique that enables power to be supplied to airborne loads in a non-contact manner. It has the advantages of electrical isolation, safety, and flexibility [1]. The technology has received extensive attention from academia and industry and is widely used in electric vehicles [2, 3], unmanned aerial vehicles (UAVs) [4–6], wireless motors [7], [8], and household equipment [9, 10]. For UAVs, since flight times are typically only about 30 minutes [11] and are in dire need of extension, WPT offers an attractive solution versus physically replacing drone batteries. The origins of WPT can be traced back to the late nineteenth century, about 100 years ago.

1.1 The Origins of Wireless Power Transfer

In the late nineteenth century, Nikola Tesla first proposed WPT technology and developed a wireless lighting system that used inductive and capacitive near-field coupling. He gave public demonstrations and lectures on the technology at Columbia College, lighting Geissler tubes and incandescent bulbs from across the stage [12]. In order to realize WPT at that time, an 'oscillating transformer' was applied to generate high-voltage, low-current, and high-frequency alternating current, which later became known as a 'Tesla coil'. Nikola Tesla observed that the transfer distance could be extended when the receiver's inductor and capacitor (LC) circuit was adjusted to resonate together on the LC circuit of the transmitter [13] – that is, resonant inductive coupling.

In 1901, Tesla built a famous tower in Shoreham, New York, known as the Walden Clifford Tower, using an investment of \$150,000. The purpose of this tower was to wirelessly transmit electricity and for use in experiments in transatlantic radio broadcasting. The project stalled by 1905 due to Tesla's financial problems and a lack of new investment.

Over the next few decades, although Tesla worked on it with the help of a number of different investors, the WPT technology was not commercialized as a full-fledged product due to the limitations of the period in terms of semiconductor materials, power electronics, and manufacturing. Although this research on WPT did not continue in the early twentieth century, Tesla opened up a whole new field of research, and thus he is known as the father of wireless power transfer [14].

After nearly a century of stagnation in WPT technology, in 2007, at the US Massachusetts Institute of Technology (MIT), Professor Marin Soljacic and his research team succeeded in lighting a 60 W electric bulb from about 2 m away, with a transmission efficiency of 40%. They used two 30 cm radius primary and pickup resonator coils and adjusted the resonant frequency of the primary circuit so that both coils resonated at 10 MHz. By exploiting this magnetic field resonance (i.e. the principle of magnetic coupling resonance), they succeeded in lighting the bulb. This method not only makes up for the short transmission distance defects of inductive non-contact WPT technology, and improves the transmission distance to the metre range, it also greatly reduces the impact of energy transmission on the environment due to its low electromagnetic radiation. These researchers published their results in *Science* in 2007 [15]; the news attracted widespread attention in the academic community and set off a new round of WPT technology research.

1.2   Inductive Power Transfer

Wireless power transfer utilizes a transmitter to generate an electromagnetic field to transmit energy via non-contact means through space, such as air, to a receiving device, thus enhancing the versatility and reliability of electronic equipment. WPT can be mainly categorized into two types based on the different transmission distances: close-field and far-field power transfer (also known as near-field and long-field power transfer, respectively).

Close-field transmission mainly consists of capacitive power transfer (CPT) and inductive power transfer (IPT). For near-field transmission, inductive coupling between coils or capacitive coupling between capacitor plates is utilized to transmit power over short distances, which can be used in electric vehicles, UAVs, and household equipment. Far-field transmission mainly includes acoustic, optical, and microwave WPT, where power is transmitted through beams of electromagnetic radiation. It is used in space technology and military applications, such as for energy transfer between space stations and spacecraft. Although a larger transfer distance can be achieved by this means, the transmitter needs to be aimed at the receiver. Currently, IPT has the most promising development and application prospects, which we explore further below.

1.2.1   Fundamentals

Researchers around the world have extensively studied the IPT system, dividing it into transmitting and receiving components, as illustrated in Fig. 1.1. The generation of a high-frequency AC signal can be achieved using the DC power supply and an inverter, which is driven using the pulse width modulation (PWM) output. By utilizing electromagnetic induction, the transmitting current is transferred from the primary coil to the coupling pickup coil without a cable connection, which provides greater safety and flexibility [16]. Employing the principle of magnetic resonance, a suitable

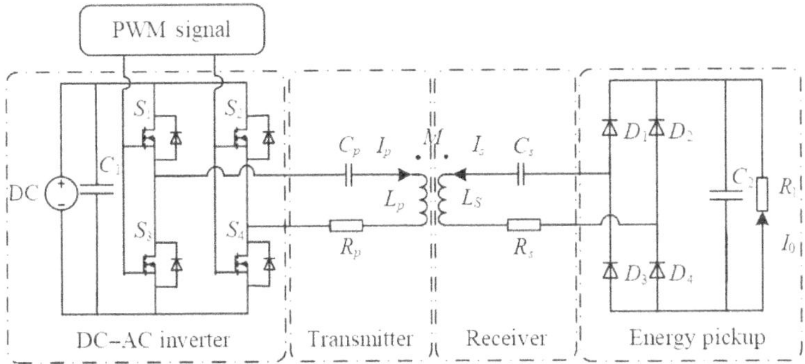

Figure 1.1 Inductive power transfer.

compensation topology is used for the primary and pickup coils so that the circuits satisfy the resonance state – that is, the magnetic coupling resonance (MCR) is realized, which ensures medium-distance WPT.

Using the principle of magnetic resonance from the physical world, in 2007 Professor Marin Soljacic of MIT designed some experiments to verify the MCR system – that is, to ensure the transmitting resonator resonates at the same frequency as the receiving resonator. He discovered that electrical energy can be transmitted between these resonators. This famous WPT experiment effectively verified the MCR mechanism at a distance of 2 m and transmission power of 60 W [15]. In order to further analyse and enhance the transfer efficiency and range of this WPT system, other scholars have expanded upon the theoretical analysis by modelling and analysing the system using equivalent circuits and Neumann's formula [17]. For a typical application of wireless energy transmission (e.g. a hovering UAV wireless charging system), a novel estimation scheme of mutual inductance has been designed based on high-order circuit topology, and a constant-current controller has been designed to ensure constant output current control without receiver-side current feedback [18]. High-order compensation topologies are employed in this context to achieve load-independent constant-current output characteristics.

In some of the literature, magnetically induced WPT systems have been categorized into two types: the IPT system and the MCR system. The IPT system is described as a type of WPT between transmitter and receiver coils in close proximity by means of magnetic coupling based on the transformer principle. The MCR WPT system ensures that the transmitting and receiving mechanisms are in a common resonance state, and utilizes the energy coupling principle to realize WPT over medium distances. In IPT systems the actual compensating capacitance is used to resonate with the coil, while MCR systems use the designed parasitic capacitor to balance and compensate for the impedance. The fundamentals and operating principles of the two categories are essentially identical, but the quality factor (Q) of the coil is usually higher in the MCR system to ensure greater coupling [19]. Therefore, instead of distinguishing between these two systems in this book, the more representative IPT system is used to illustrate the magnetic induction-based WPT system.

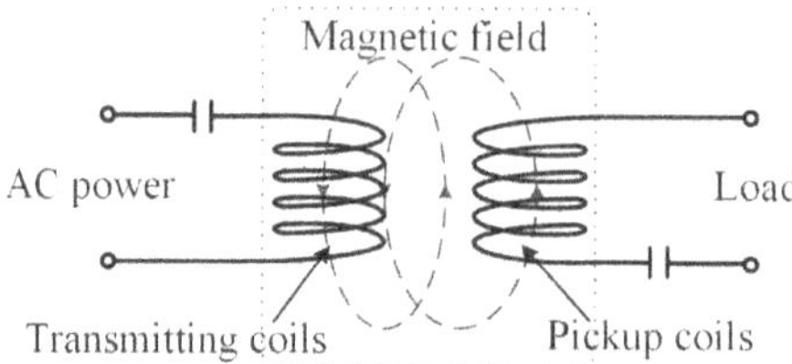

Figure 1.2 The principle of IPT.

Figure 1.2 shows the principle behind the IPT system, where the AC voltage is supplied to the load side in a non-contact manner through the transmitting and receiving coils. The specific power transfer process is described as follows:

1. A high-frequency inverter is used to convert the DC voltage generated from a DC power source into a high-frequency AC voltage (from tens of kilohertz [20, 21] to a few megahertz [22]), which is the input to the WPT system.
2. The AC voltage is input to the coil inductor and compensation capacitor of the primary side at resonance, while the alternating current flowing through the transmitting coil produces an alternating induced magnetic field.
3. The alternating induced magnetic field of a given frequency produced by the transmitting coil induces an alternating inductive voltage of the identical frequency in the receiving coil through weak coupling.
4. After capacitive compensation, rectification, and filtering on the receiving side, the induced AC voltage is transformed into a DC voltage and transmitted to the power-using equipment.

From the above elaboration, it can be seen that the principle of the IPT system is familiar to traditional transformer theory. So what are the differences between them? In the following, we discuss four important characteristics of IPT systems that differ from transformers.

Lesser Coupling Coefficients

Due to the existence of a transfer distance between the two coils in this IPT system, the resulting leakage inductance is generally hundreds of times greater than that of a transformer. The larger air gap also greatly reduces the magnetic flux, resulting in a lower mutual inductance coefficient M and a weaker coupling coefficient in this IPT system [23]. Therefore, to obtain higher Q coils and enhance the system coupling coefficient, the optimized design of the coupling coils and the reasonable selection of compensation topology are key difficulties in the IPT system.

Hollow-Core Inductor

In IPT systems, power transfer between the primary and pickup devices can be realized by magnetic coupling between inductors, most of which are hollow inductors in order to avoid core-induced energy loss. The difference is that transformer windings usually use cores to form the magnetic circuit of the system to minimize energy loss due to magnetic leakage. Therefore, whether or not a core is used as the magnetic circuit can be a critical factor in distinguishing a conventional transformer from an IPT system.

Phase Difference

For an ideal transformer, if we ignore losses such as leakage inductance, the transmitter and receiver currents may be the equal or reverse of each other; for an IPT system the phase difference of the transmitter- and receiver-side currents can usually be about 90 degrees because of the impedance adjustment of the compensation capacitors.

System Disturbances and Parasitic Parameters

As the transfer distance between coils, angular/lateral misalignment, and load values fluctuate, the intrinsic and reflective values of an IPT system vary dramatically, resulting in dynamic changes in output power and efficiency. Furthermore, in common high-frequency IPT systems it is important to note that parasitic resistors and parasitic capacitors have a significant impact on the system transfer capability. Therefore, how to maintain great transfer efficiency and robust output behaviour in the presence of complicated parametric disturbances and parasitic parameters is a key challenge for IPT systems.

This section has provided a comprehensive overview of the IPT principle, including the system components and the differences in the principle relative to conventional transformers. The general configuration of IPT systems is described next.

1.2.2 General Configuration

As shown in Fig. 1.3, a common IPT system configuration uses only one transmitter and one receiver, with typical components including: primary-side energy supply, transmitter-side compensated topology, magnetic coupler, receiver-side compensated topology, and receiver device.

Primary-Side Power Source

An inverter is the key component for converting DC voltage into the high-frequency AC voltage used in IPT systems. Half- or full-bridge class D inverters have been widely used in low- and medium-power IPT systems for their simple parameter setup

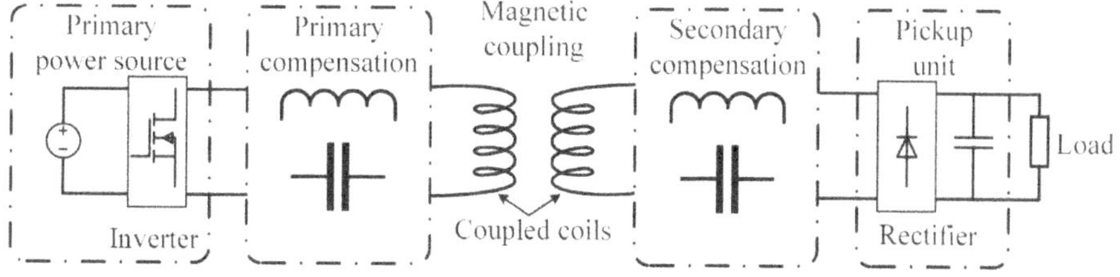

Figure 1.3 Configuration of an IPT system.

and ease of control. However, since the surge voltage flowing through the switching tubes is generally double that of the input voltage, the switching tubes face the risk of breakdown at high voltages. To address this problem, the parallel and cascaded inverters can be adopted to achieve increased output power with improved flexibility [24]. It is worth noting that the combination of signal generator and power amplifier can also generate an AC input voltage to power the IPT system; this has lower total harmonic distortion (THD) but poorer efficiency performance.

Primary/Secondary-Side Compensation Topology
A specially designed compensation topology can eliminate the large leakage inductive reactance in the IPT loosely coupled system by means of several LC resonant tanks, minimizing the primary-side voltage-ampere (VA) rating and maximizing the receiver-side efficiency [25, 26]. In the primary and secondary resonant circuits, the coil inductances resonate with the capacitances of the capacitors. Meanwhile, the LC resonant circuit filters out harmonics in other frequency bands and modulates the transmitter current to a sinusoidal waveform.

Common compensation topologies are series or parallel compensation, and some high-order topologies such as the LCC topology can be used to achieve a constant load-independent output under parametric disturbances, or to implement soft-switching techniques to maximize the transmission efficiency of the IPT system. In high-frequency IPT systems, MOSFETs (metal-oxide-semiconductor field-effect transistors) are often used for inverter switching components, and the implementation of soft-switching techniques can effectively reduce switching power loss and thus greatly improve the overall transfer efficiency.

In conclusion, a proper selection of the compensation network and the optimal parameter design of the LC compensation topology can greatly benefit the transfer properties of the IPT system.

Magnetic Coupling
Figure 1.4 illustrates a simplified solenoidal magnetic coupling system in an IPT system. Here, r is the coil radius and d is the transmission distance. According to the Biot–Savart theorem, the strength of the electromagnetic force at a point is:

$$H(d) = \frac{IN}{2} \frac{r^2}{\left(r^2 + d^2\right)^{\frac{3}{2}}}.$$ (1.1)

Based Eq. (1.1), the magnetic field strength is enhanced along with the increase of the primary current I or the increase of the number of turns N of the coil. Theoretically, the higher the magnetic field strength of the WPT system, the higher the output power, but the calculation of r, d, and the magnetic field strength is complicated. In the plot shown in Fig. 1.5, the horizontal coordinate is the ratio of d to r and the vertical coordinate is the normalized magnetic field strength. It can be seen that the normalized magnetic field strength increases slowly and then decreases sharply as the horizontal coordinate d/r increases. Based on Eq. (1.1), it can be found that at the peak of the

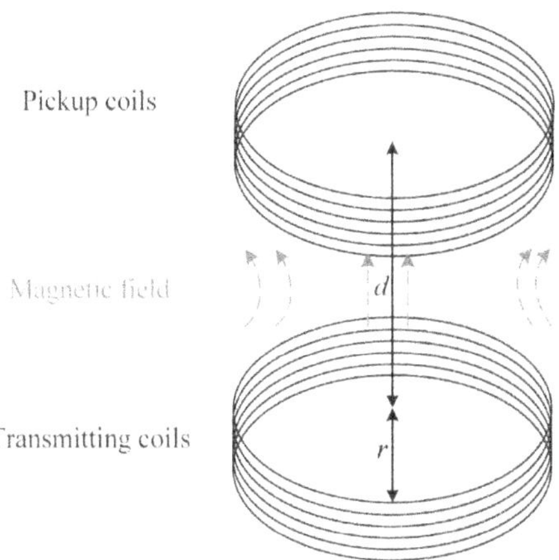

Figure 1.4 A simplified magnetic coupling system.

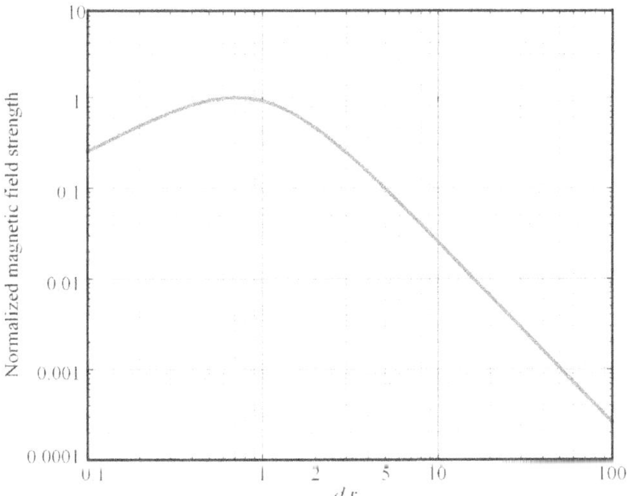

Figure 1.5 The system magnetic field strength.

magnetic field, the ratio of transmission distance d to radius r is 1/sqrt(2). Therefore, transmission performance is better when the transfer distance is the radius of the transmitting coil.

Receiver Unit

Generally, a rectifier bridge is used to transform the AC current to DC current on the receiving side, followed by a filter capacitor to reduce the voltage ripple. In addition, the DC–DC converter may be required to stabilize the load voltage or current, depending on the required load voltage or current of the power-using equipment. The converter can also be adjusted with a switching duty cycle to be applied as an impedance transformer.

1.3   Theoretical Modelling

Figure 1.3 shows the structure of a single-transmitter and single-receiver IPT system, where the loosely coupled transmitter-side and receiver-side coils are not mechanically or electrically connected, but rather the power is transmitted through an induced magnetic field. Therefore, magnetic coupling is the key component of the IPT system required to realize power transmission. Here, we begin by describing the basic fundamentals of loosely coupled transformers.

1.3.1   The Model for Loosely Coupled Transformers

For practical loosely coupled transformer systems, a large leakage inductance and leakage flux are generated because of the long transfer distance of the coupling coils. Figure 1.6 shows a coupled transformer with leakage flux, where the flux distribution of the coupling coils is depicted.

In Figure 1.6 ϕ_{ps} represents the flux produced by the primary coil passing through the pickup coil, and ϕ_{pl} represents the flux passing elsewhere; similarly, ϕ_{sp} represents the flux generated by the pickup coil passing through the primary coil and ϕ_{sl} represents the flux passing elsewhere. In addition, U_p, I_p, U_s, and I_s represent the primary current voltage and the secondary current voltage, respectively; ϕ_m represents the shared flux:

$$\phi_m = \phi_{ps} - \phi_{sp}. \tag{1.2}$$

In this case, the equivalent magnetic flux produced in the primary coil is:

$$\phi_p' = \phi_{ps} - \phi_{sp} + \phi_{pl}. \tag{1.3}$$

In a similar way, the equivalent magnetic flux produced in the pickup coil is:

$$\phi_s' = \phi_{sp} - \phi_{ps} + \phi_{sl}. \tag{1.4}$$

Therefore, the primary-side input voltage and charging voltage are:

$$
\begin{aligned}
U_p &= N_p \frac{d\phi_p'}{dt} = N_p \frac{d\left(\phi_{pl} + \phi_{ps} - \phi_{sp}\right)}{dt} = L_{pl}\overrightarrow{i_p} + \left(L_{ps}\overrightarrow{i_p} - \frac{L_{sp}\overrightarrow{i_s}}{n}\right), \\
U_s &= N_s \frac{d\phi_s'}{dt} = N_s \frac{d\left(\phi_{sl} + \phi_{sp} - \phi_{ps}\right)}{dt} = L_{sl}\overrightarrow{i_s} + \left(L_{sp}\overrightarrow{i_s} - nL_{ps}\overrightarrow{i_p}\right).
\end{aligned}
\tag{1.5}
$$

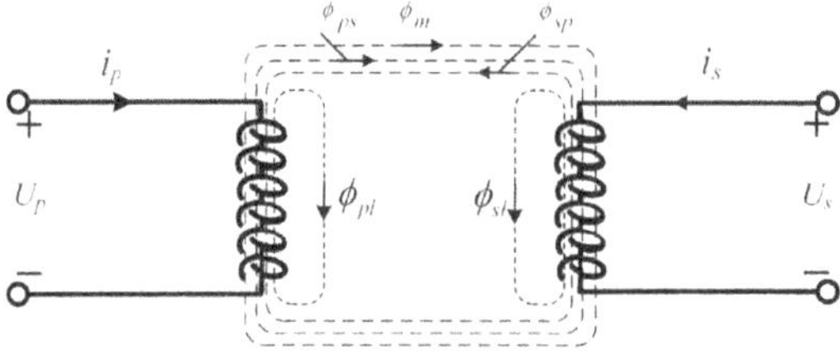

Figure 1.6 A schematic diagram of loosely coupled transformers.

Here, N_p and N_s are the turns of the primary and secondary windings. In addition, the relative inductance and turns ratio n can be indicated as:

$$L_{pl} = \frac{N_p \phi_{pl}}{i_p}, L_{sl} = \frac{N_s \phi_{sl}}{i_s}, L_{ps} = \frac{N_p \phi_{ps}}{i_p}, L_{sp} \frac{N_s \phi_{sp}}{i_s}, n = \frac{N_s}{N_p}. \tag{1.6}$$

In this case, the equation of the transformer magnetic circuit is expressed in the following way:

$$N_p i_p = N_s i_s = \mathfrak{R} \phi_m, \tag{1.7}$$

where $\mathfrak{R}$ is the magnetoresistance of the coil. Following this, L_{ps} can be recalculated as:

$$L_{ps} = \frac{N_p \phi_{ps}}{i_p} = \frac{N_p \phi_m}{i_p} \bigg|_{i_s=0}. \tag{1.8}$$

By substituting Eq. (1.7) into Eq. (1.8), L_{ps} is expressed as:

$$L_{ps} = \frac{N_p^2}{\mathfrak{R}}. \tag{1.9}$$

Similarly, L_{sp} can be derived as:

$$L_{sp} = \frac{N_s^2}{\mathfrak{R}}. \tag{1.10}$$

According to Eqs. (1.9) and (1.10), the relationship of L_{sp} and L_{ps} can be expressed as:

$$L_{sp} = n^2 L_{ps}. \tag{1.11}$$

By substituting Eq. (1.11) into Eq. (1.5), it can be concluded that:

$$\begin{aligned}
U_p &= L_{pl} \vec{i_p} + \left(L_{ps} \vec{i_p} - \frac{L_{sp} \vec{i_s}}{n} \right) = L_{pl} \vec{i_p} + L_{ps} \left(\vec{i_p} - n \vec{i_s} \right), \\
U_s &= L_{sl} \vec{i_s} + \left(L_{sp} \vec{i_s} - n L_{ps} \vec{i_p} \right) = L_{sl} \vec{i_s} - n L_{ps} \left(\vec{i_p} - n \vec{i_s} \right), \\
U_m &= L_{ps} \left(\vec{i_p} - n \vec{i_s} \right).
\end{aligned} \tag{1.12}$$

Therefore, the circuit in Fig. 1.7 is constructed according to Eq. (1.12), which is composed of an ideal transformer.

1.3.2   T-model

The T-model is illustrated in Fig. 1.8. It is called the T-model because the three inductors in its topology are similar to the letter T. The following equation is satisfied: $M = n L_{ps}$, $L_p = L_{pl} + L_{ps}$, $L_s = L_{sl} + L_{sp}$. By replacing Eq. (1.11) into Eq. (1.12), we obtain:

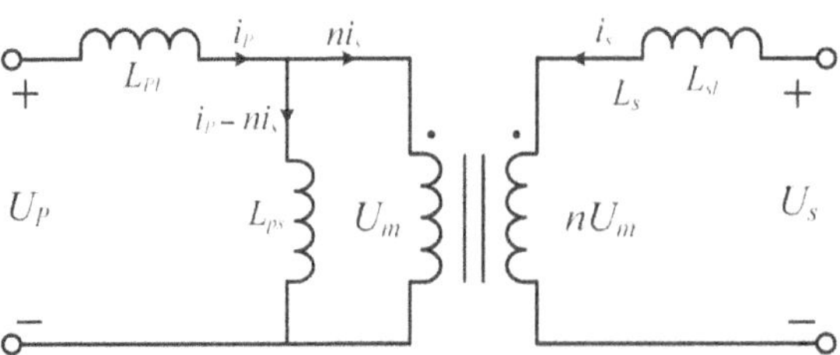

Figure 1.7 A loosely coupled transformer model.

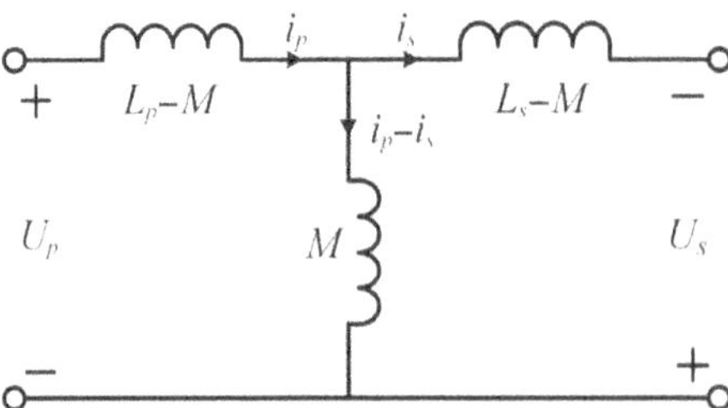

Figure 1.8 The T-model for one-to-one transmission.

$$U_p = L_{pl}\,\vec{i_p} + L_{ps}\left(\vec{i_p} - n\,\vec{i_s}\right) = \left(L_{pl} + L_{ps} - nL_{ps}\right)\vec{i_p} + nL_{ps}\left(\vec{i_p} - \vec{i_s}\right)$$

$$= \left(L_p - M\right)\vec{i_p} + M\left(\vec{i_p} - \vec{i_s}\right),$$

$$U_s = L_{sl}\,\vec{i_s} + \left(L_{sp}\,\vec{i_s} - nL_{ps}\,\vec{i_p}\right) = \left(L_{sl} + L_{sp} - nL_{ps}\right)\vec{i_s} - nL_{ps}\left(\vec{i_p} - \vec{i_s}\right)$$

$$= \left(L_s - M\right)\vec{i_s} - M\left(\vec{i_p} - \vec{i_s}\right). \tag{1.13}$$

1.3.3 M-model

The M-model (Fig. 1.9) is a common theoretical model applied to IPT systems. According to Eq. (1.13), the theoretical formulation of the M-model is derived as:

$$U_p = \left(L_p - M\right)\vec{i_p} + M\left(\vec{i_p} - \vec{i_s}\right) = L_p\,\vec{i_p} - M\,\vec{i_s},$$

$$U_s = \left(L_s - M\right)\vec{i_s} - M\left(\vec{i_p} - \vec{i_s}\right) = L_s\,\vec{i_s} - M\,\vec{i_p}, \tag{1.14}$$

where M represents the mutual inductance of coils:

$$M = nL_{ps}. \tag{1.15}$$

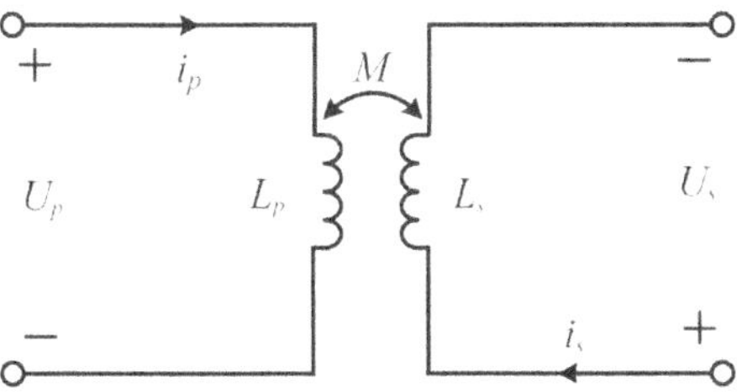

Figure 1.9 M-model for one-to-one transmission.

1.4 Compensation Network

Because of the high-leakage inductive reactance of IPT systems, without matching the compensation capacitors both the transmitter and receiver sides will consume a lot of power, thus reducing the system's power factor. Therefore, reactive power compensation is extremely important to improve the system transfer efficiency.

In general, capacitive compensation is a common method of removing inductive reactance in IPT systems. It is based on the principle of utilizing capacitors in a well-designed compensation network to resonate with the coil self-inductance, hence eliminating the system reactive power. Typically, the compensation circuits can be categorized into two forms, series type and parallel type, based on the connection between the capacitors and inductors.

1.4.1 Series Type

For the RLC circuit illustrated in Fig. 1.10(a), the equivalent impedance Z_s is:

$$Z_s = R + j\omega L + \frac{1}{j\omega C} = R + j\left(\omega L - \frac{1}{\omega C}\right) = R + jX = |Z|\angle\psi, \tag{1.16}$$

where X is the imaginary part of Z_s. Setting $X = 0$, ω_0 is determined as:

$$\omega_0 = \frac{1}{\sqrt{LC}}. \tag{1.17}$$

f_0 can be expressed as:

$$f_0 = \frac{1}{2\pi\sqrt{LC}}. \tag{1.18}$$

According to Eqs. (1.17) and (1.18), the series resonance is characterized as follows. The voltage across the inductor ($U_{L0} = j\omega_0 L \overrightarrow{I_0}$) and the voltage across the capacitor ($U_{C0} = -j\omega_0 \frac{1}{C_0} \overrightarrow{I_0}$) have the same amplitude and opposite phase. Thus, the

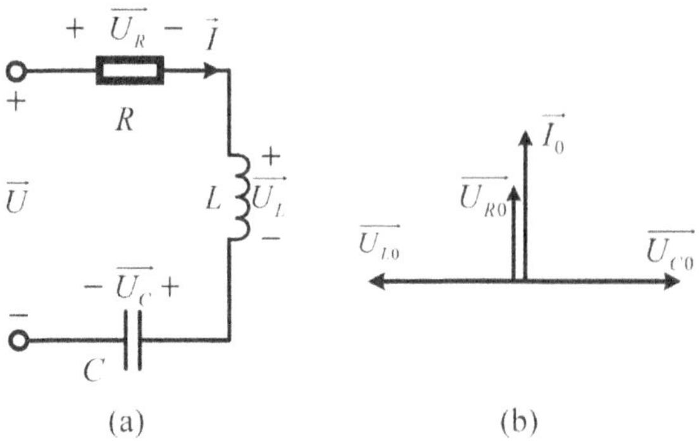

(a)   (b)

Figure 1.10 (a) Series-type circuit; (b) series resonance vector.

LC resonant tank is equivalent to a short circuit and all voltages are supplied to the resistor, as shown in Fig. 1.10(b).

The quality factor (Q) of a resonant circuit can be defined by the ratio of the resonant inductor voltage or resonant capacitor voltage to the supply voltage:

$$Q = \frac{U_{L0}}{U} = \frac{U_{C0}}{U} = \frac{\omega_0 L}{R}. \tag{1.19}$$

1.4.2 Parallel Type

The IPT system with parallel capacitor compensation is illustrated in Fig. 1.11(a). Z_p is expressed as:

$$Z_p = \frac{(R + j\omega L)\frac{1}{j\omega C}}{R + j\omega L + \frac{1}{j\omega C}}. \tag{1.20}$$

Assuming $\text{Im}\{Z_p\} = 0$, ω_0 is obtained by:

$$\omega_0 = \sqrt{\frac{1}{LC} - \frac{R^2}{L^2}}. \tag{1.21}$$

Equation (1.21) needs to satisfy $R < \sqrt{\frac{L}{C}}$. f_0 can be expressed as:

$$f_0 = \frac{1}{2\pi}\sqrt{\frac{1}{LC} - \frac{R^2}{L^2}}. \tag{1.22}$$

Figure 1.11(b) illustrates the parallel-type vector map. Some higher-order topologies have been applied to achieve the desired output characteristics, such as constant current.

Because of the advantages of the simplicity of the structure and independence of the parameter design, this chapter focuses on the primary-side series–secondary-side series (SS) compensation network for the transmission performance.

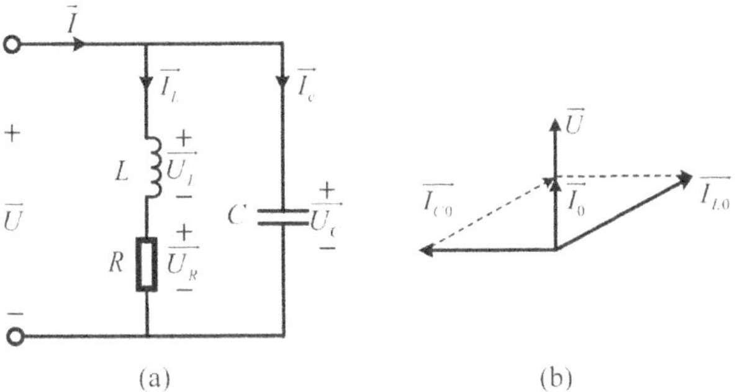

Figure 1.11 (a) Parallel-type circuit; (b) parallel resonance vector.

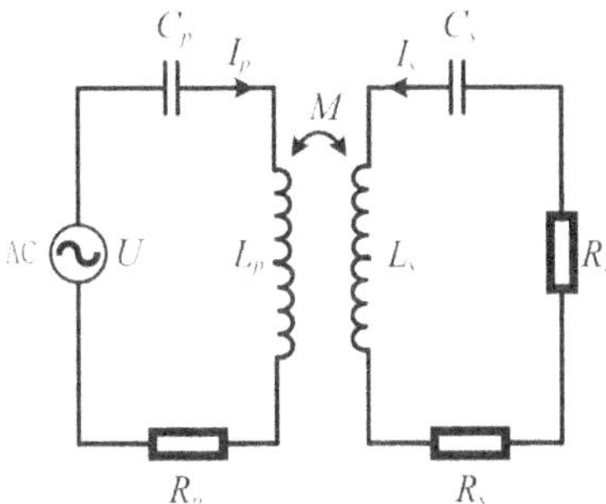

Figure 1.12 A simplified IPT system with SS compensation network.

1.5 Transmission Performance

This section analyses the transmission performance of the most common SS compensated network for single-transmitter–single-receiver IPT systems, focusing on the output power, output efficiency, and their influencing factors.

1.5.1 Output Power

The IPT system in SS topology is shown in Fig 1.12. Its circuit equations are:

$$\begin{bmatrix} \vec{U} \\ 0 \end{bmatrix} = \begin{bmatrix} R_p + j\omega L_p + \dfrac{1}{j\omega C_p} & M_{ps} \\ M_{ps} & R_L + R_s + j\omega L_s + \dfrac{1}{j\omega C_s} \end{bmatrix} \begin{bmatrix} \vec{I_p} \\ \vec{I_s} \end{bmatrix}, \qquad (1.23)$$

where L, C, and R, subscripted as p or s, are the transmitting and receiving inductance, compensation capacitance, and coil internal resistance, respectively. The system load and mutual inductance between the coupling coils are $\underline{R_L}$ and M_{ps}, respectively.

When the system impedance is matched, Eq. (1.24) can be derived:

$$j\omega L_k + \frac{1}{j\omega C_k} = 0, k \in [p,s].\tag{1.24}$$

Replacing Eq. (1.24) into Eq. (1.23), the current I_p and $\overrightarrow{I_s}$ are derived as:

$$I_p = \frac{U}{R_p + \dfrac{\omega^2 M_{ps}^2}{R_s + R_L}} = \frac{U}{Z_{in}},$$

$$\overrightarrow{I_s} = \frac{-j\omega M_{ps} U}{R_p(R_s + R_L) + \omega^2 M_{ps}^2}.\tag{1.25}$$

The reflected impedance Z_r from the receiver to the transmitter can be calculated as:

$$Z_r = \frac{\omega^2 M_{ps}^2}{R_s + R_L}.\tag{1.26}$$

According to Eq. (1.25), the input power P_{in} and the output power P_L are derived as:

$$P_{in} = \frac{(R_s + R_L)U^2}{R_p(R_s + R_L) + \omega^2 M_{ps}^2},$$

$$P_L = \frac{\omega^2 M_{ps}^2 U^2 R_L}{\left[R_p(R_s + R_L) + \omega^2 M_{ps}^2\right]^2}.\tag{1.27}$$

Here, the optimal load R_{Lpmax} of the largest output power is obtained as:

$$R_{Lpmax} = \frac{\omega^2 M_{ps}}{R_p} + R_s.\tag{1.28}$$

Then, by substituting Eq. (1.28) into Eq. (1.27), the largest output power P_{Lmax} is obtained:

$$P_{Lmax} = \frac{\omega^2 M_{ps}^2 U^2}{4\left(\omega^2 M_{ps}^2 + R_p R_{s1}\right) R_p}.\tag{1.29}$$

1.5.2   Transfer Efficiency

It is worth noting that the transfer efficiency of an IPT system is one of its key metrics. According to Eq. (1.27), the transfer efficiency η is obtained as:

$$\eta = \frac{P_L}{P_{in}} = \frac{\omega^2 M_{ps}^2 R_L}{\left[R_p(R_s + R_L) + \omega^2 M_{ps}^2\right](R_s + R_L)}.\tag{1.30}$$

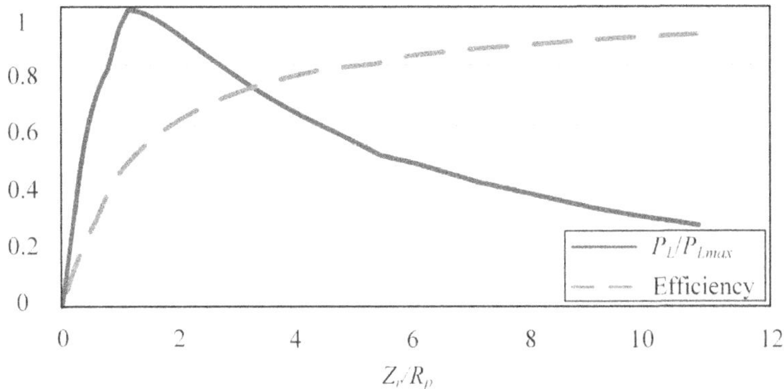

Figure 1.13 Efficiency and normalized load power versus Z_r/R_p.

The optimal load $R_{L\eta max}$ of the greatest efficiency is obtained as:

$$R_{L\eta max} = R_{s1}\sqrt{1 + \frac{\omega^2 M_{ps}^2}{R_p R_s}}. \tag{1.31}$$

Then, based on Eqs. (1.31) and (1.30), the greatest efficiency is obtained as:

$$\eta_{max} = 1 - \frac{2}{\sqrt{1 + \frac{\omega^2 M_{ps}^2}{R_p R_s}} + 1}. \tag{1.32}$$

Therefore, according to Eq. (1.32), it can be found that to obtain a larger transfer efficiency, it is theoretically possible to reduce the coil internal resistance and increase the coupling effect of the coils.

1.5.3   The Relationship between Output Power and System Efficiency

To further illustrate the coupling between output power and system efficiency, Fig. 1.13 shows the efficiency and normalized load power against Z_r/R_p. It can be seen that the normalized load power initially rises rapidly then falls slightly with increasing values of Z_r/R_p. It is worth noting that the maximum output power is reached when $Z_r/R_p = 1$ is satisfied. In addition, the transfer efficiency rises non-linearly as Z_r/R_p increases, eventually approaching 1.

References

[1] Z. Zhang, H. Pang, A. Georgiadis, and C. Cecati, 'Wireless power transfer: an overview', *IEEE Trans. Ind. Electron.*, vol. 66, no. 2, pp. 1044–1058, 2019.

[2] S. Li, S. Lu, and C. C. Mi, 'Revolution of electric vehicle charging technologies accelerated by wide bandgap devices', *Proc. IEEE*, vol. 109, no. 6, pp. 985–1003, Jun. 2021.

[3] X. Dai, J. Jiang, and J. Wu, 'Charging area determining and power enhancement method for multiexcitation unit configuration of wirelessly dynamic charging EV system', *IEEE Trans. Ind. Electron.*, vol. 66, no. 5, pp. 4086–4096, May 2019.

[4] Y. Gu, J. Wang, Z. Liang, and Z. Zhang, 'A wireless in-flight charging range extended PT-WPT system using S/single-inductor-double-capacitor compensation network for drones', *IEEE Trans. Power Electron.*, vol. 38, no. 10, pp. 11847–11858, Oct. 2023.

[5] W. Han, K. T. Chau, C. Jiang, W. Liu, and W. H. Lam, 'Design and analysis of quasi-omnidirectional dynamic wireless power transfer for fly-and-charge', *IEEE Trans. Magn.*, vol. 55, no. 7, pp. 1–9, Jul. 2019.

[6] Y. Gu, J. Wang, Z. Liang, and Z. Zhang, 'Communication-free power control algorithm for drone wireless in-flight charging under dual-disturbance of mutual inductance and load', *IEEE Trans. Ind. Inform.*, vol. 20, no. 3, pp. 3703–3714, Mar. 2024.

[7] C. Jiang, K. T. Chau, C. H. T. Lee, W. Han, W. Liu, and W. H. Lam, 'A wireless servo motor drive with bidirectional motion capability', *IEEE Trans. Power Electron.*, vol. 34, no. 12, pp. 12001–12010, Dec. 2019.

[8] W. Han, K. T. Chau, Z. Hua, and H. Pang, 'An integrated wireless motor system using laminated magnetic coupler and commutative-resonant control', *IEEE Trans. Ind. Electron.*, vol. 69, no. 5, pp. 4342–4352, May 2022.

[9] Y. Gu, J. Wang, Z. Liang, Y. Wu, C. Cecati, and Z. Zhang, 'Single-transmitter multiple-pickup wireless power transfer: advantages, challenges, and corresponding technical solutions', *IEEE Ind. Electron. Mag.*, vol. 14, no. 4, pp. 123–135, Dec. 2020.

[10] X. Qu, W. Zhang, S.-C Wong, and C. K. Tse, 'Design of a current-source-output inductive power transfer LED lighting system', *IEEE J. Emerg. Sel. Topics Power Electron.*, vol. 3, no. 1, pp. 306–314, Mar. 2015.

[11] Z. Zhang, S. Shen, Z. Liang, S. H. K. Eder, and R. Kennel, 'Dynamic-balancing robust current control for wireless drone-in-flight charging', *IEEE Trans. Power Electron.*, vol. 37, no. 3, pp. 3626–3635, Mar. 2022.

[12] Wikipedia. 'Wireless power transfer'. Available: https://en.wikipedia.org/wiki/Wireless_power_transfer.

[13] L. P. Wheeler, 'Tesla's contribution to high frequency', *Electr. Eng.*, vol. 62, no. 8, pp. 355–357, 1943.

[14] N. Tesla, 'System of Transmission of Electrical Energy', U.S. Patent 645,576, Mar 20, 1900.

[15] A. Kurs, A. Karalis, R. Moffatt, J. D. Joannopoulos, P. Fisher, and M. Soljacic, 'Wireless power transfer via strongly coupled magnetic resonances', *Science*, vol. 317, no. 5834, pp. 83–86, Jul. 2007.

[16] G. A. Govic and J. T. Boys, 'Inductive power transfer', *Proc. IEEE*, vol. 101, no. 6, pp. 1276–1289, 2013.

[17] T. Imura and Y. Hori, 'Maximizing air gap and efficiency of magnetic resonant coupling for wireless power transfer using equivalent circuit and Neumann formula', *IEEE Trans. Ind. Electron.*, vol. 58, no. 10, pp. 4746–4752, 2011.

[18] Y. Gu, J. Wang, Z. Liang, and Z. Zhang, 'Mutual-inductance-dynamic-predicted constant current control of LCC-P compensation network for drone wireless in-flight charging', *IEEE Trans. Ind. Electron.*, vol. 69, no. 12, pp. 12710–12719, Dec. 2022.

[19] C.-J. Chen, T.-H. Chu, C.-L. Lin, and Z.-C. Jou, 'A study of loosely coupled coils for wireless power transfer', *IEEE Trans. Circuits Syst. II*, vol. 57, no. 7, pp. 536–540, 2010.

[20] J. Huh, S. W. Lee, W. Y. Lee, G. H. Cho, and C. T. Rim, 'Narrow-width inductive power transfer system for online electrical vehicles', *IEEE Trans. Power Electron.*, vol. 26, no. 12, pp. 3666–3679, 2011.

[21] PMA inductive wireless power and charging transmitter specification-system release 1, Standard PMA-TS-003-0 v1.00, 2014.

[22] A4WP wireless power transfer system baseline system specification, BSS Standard A4WP-S-0001 v1.2, 2014.

[23] J. O. Mur-Miranda, G. Fanti, Y. Feng, K. Omanakuttan, R. Ongie, A. Setjoadi, and N. Sharpe, 'Wireless power transfer using weakly coupled magnetostatic resonators', in *IEEE Ener. Conv. Congr. Expos.*, 2010, pp. 4179–4186.

[24] C. Jiang, K. T. Chau, C. Liu, and C. H. T. Lee, 'An overview of resonant circuits for wireless power transfer', *Energies*, vol. 10, no. 7, pp. 894–913, 2017.

[25] S. Hui, 'Magnetic resonance for wireless power transfer [a look back]', *IEEE Power Electron. Mag.*, vol. 3, no. 1, pp. 14–31, 2016.

[26] W. Zhang and C. C. Mi, 'Compensation topologies of high-power wireless power transfer systems', *IEEE Trans. Vehic. Technol.*, vol. 65, no. 6, pp. 4768–4778, 2016.

2 Hovering Wireless Charging for UAVs

Compared with wired charging, wireless power transfer (WPT) can efficiently supply electricity to unmanned aerial vehicles (UAVs) because of its flexibility, safety, and mobility. Wireless charging for UAVs includes both wireless charging by landing at a fixed position and wireless in-flight charging. Wireless in-flight charging of UAVs enables them to remain airborne and perform continuous detection tasks indefinitely, without the need for a mid-flight landing. This chapter details the advantages, challenges, and the latest research advances in wireless in-flight charging for UAVs.

2.1 Introduction

Unmanned aerial vehicles have broad application uses in commercial and military fields due to their compact and flexible characteristics. They can replace human beings in performing many difficult aerial tasks, including aerial photography and inspection. There has been a boom in the demand for UAVs since 2017, with a high average yearly growth rate of 23.5% up to 2021 [1]. New types of UAVs – electric-powered multi-rotor systems – have been keenly developed for practical applications [2]. By integrating new smart techniques, such as image processing and artificial intelligence, UAVs have demonstrated broader application prospects.

Market statistics indicate a clear upward trend in the use of industrial UAVs, where UAVs are replacing humans in the areas of safety management, aerial surveys, and daily inspections, owing to their flexibility, safety, and high efficiency (Fig. 2.1(a)) [3]. Nevertheless, due to the limitations of on-board battery capacity, the average maximum flight time of an electrically driven UAV is about 40 minutes, and this relatively low endurance makes it difficult to satisfy current mission demands for lengthy and wide-area continuous operations [4, 5], which limits their applications and use in industrial scenarios.

Two main approaches to these problems have been proposed in previous research. The first one is to utilize current advanced battery materials to increase battery capacity, but this may increase the weight of the battery, and so make it difficult to effectively increase the range of the drone. A second option is wired and wireless charging technologies to provide energy to drones. Wireless charging technologies allow greater flexibility and safety compared to traditional charging methods using cables.

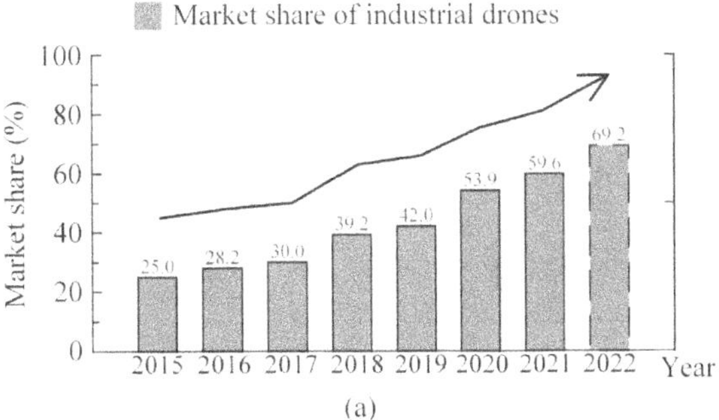

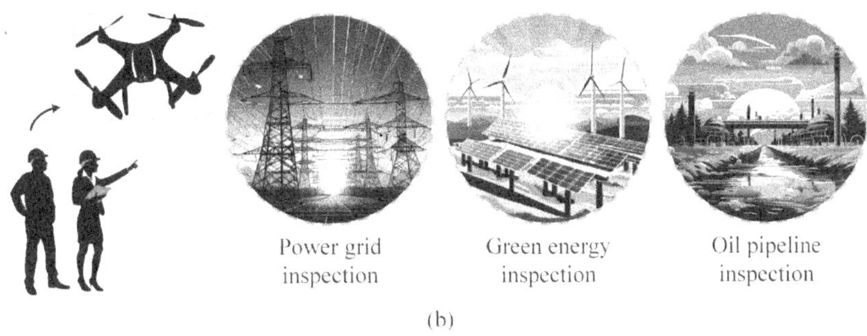

Figure 2.1 (a) The UAV market; (b) UAV applications.

As a breakthrough technology, WPT promises to become a key solution to the battery problem. In recent years, this novel power transfer method has been widely noticed and explored by researchers and technicians [6–8]. At the same time, increasing numbers of devices employing inductive WPT technology are entering our daily lives – for example, electric vehicles [9, 10], smart homes [11, 12], and implantable medical devices [13, 14].

For UAVs in particular, considering the need to use them for continuous operations over large ranges, wireless in-flight charging is an attractive option to address the major constraint of short battery lives. Figure 2.2 shows how a hovering wireless charging system could replenish the electrical energy of in-flight UAVs through inductive coupling between coils. Figure 2.3 offers a comparison of power options – battery capacity, battery replacement, and docked wireless charging – to show why WPT would be particularly useful in this case.

Option 1. An increase in the on-board battery capacity is bound to increase the UAV's load, which leads to additional power loss and fails to significantly improve the UAV's durability.

Option 2. Timely replacement of the UAV battery makes it difficult to realize intelligent and unmanned operation, and the startups and shutdowns of UAVs will also reduce their range.

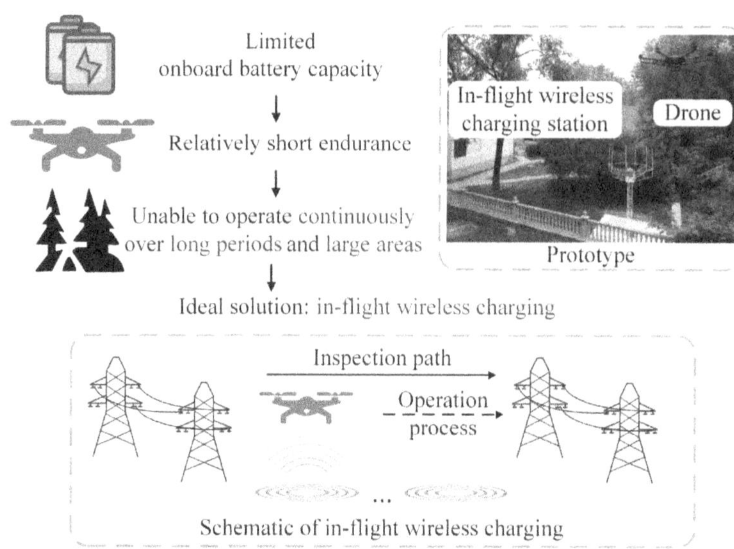

Figure 2.2 Schematic diagram of a UAV hover charging system.

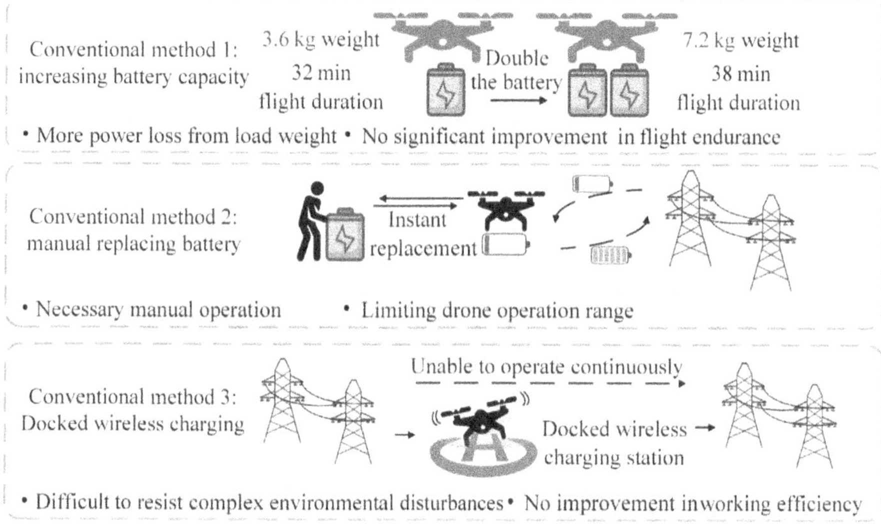

Figure 2.3 A comparison of different technology solutions.

Option 3. The installation of a docking wireless charging station can provide energy
for the UAV, but because of the shutdown of the flight control system it is not
possible to guarantee reliable energy replenishment under complicated
environmental situations such as strong wind. Furthermore, in specific industrial
situations such as large-area inspection, the docking charging scheme will

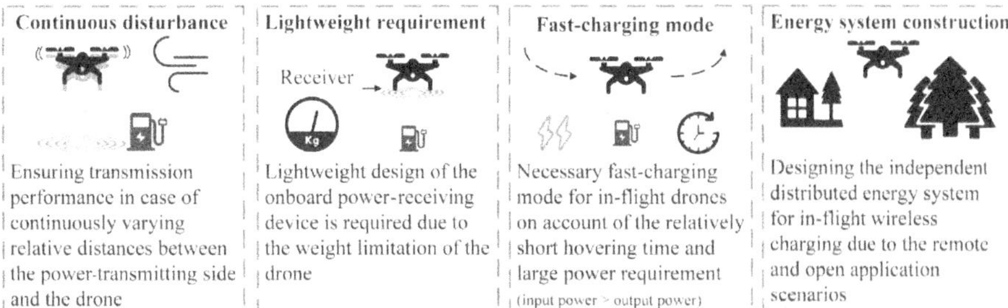

Figure 2.4 Key issues for hover charging.

undoubtedly disrupt the tasks carried out by the UAV, which is not conducive to improved work efficiency.

In summary, wireless in-flight charging technology may be an ideal solution for replenishing power to UAVs, as it can significantly prolong their operating time and range, realizing continuous operation with effective power transfer. This emerging solution for industrial applications still faces technical challenges that are different from those of other applications (Fig. 2.4), so analysing the key issues of wireless in-flight charging systems and designing corresponding solutions is key to implementing this emerging technology.

2.2 The Advantages of Hovering Wireless Charging

Common UAV wireless charging methods include wireless charging after landing on a platform [15] and hovering wireless charging [16–18].

For wireless charging after landing on a deck, an intelligent UAV charging substation has been proposed [19], where the UAV lands and is wirelessly charged autonomously, with no manual assistance. Figure 2.5 depicts the WPT system architecture, where the receiving coil and circuits are installed on the landing frame, a three-dimensional aluminium tube that is shaped to fit both mechanical and electrical features. This design reduces the weight of the UAV and has minimal influence on its aerodynamics. With the receiving coils identified, the transmitter coil is optimally designed to ensure high transmission efficiency and high misalignment resistance of the system, which includes the specific coil design scheme [19]. Landing the UAV on a particular spot successfully and safely is the main concern in this landing–charging mode.

One reason that hovering wireless charging of UAVs is gaining interest in academia and business is that it avoids the issue of system failure due to an inability to land the UAV. Hover charging allows the UAV to hover above the charging stage and pick up the power from the transmitter in a non-contact manner. Figure 2.6 shows how the UAV carrying the receiving coil hovers to charge wirelessly, and how the UAV can dynamically regulate its position to maintain high coupling strength when hovering.

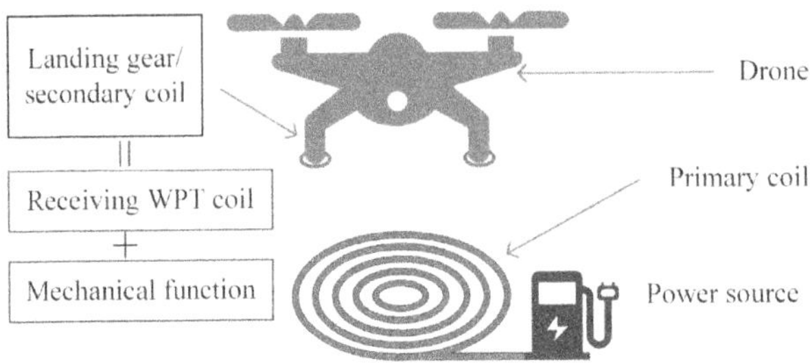

Figure 2.5 Wireless charging a UAV after landing on the platform.

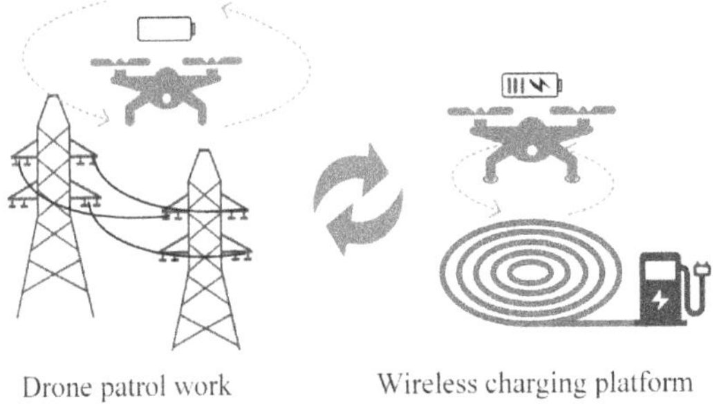

Figure 2.6 Diagram of a UAV hovering wireless charging system.

Figure 2.7 compares hover charging with landed charging, and shows where hover charging may offer advantages. The main differences between these two technologies are:

1. Landed wireless charging has a safety risk, as during the charging process the UAV flight control system is turned off, which makes it harder to resist harsh environmental influences (e.g. wind). Hovered charging keeps the flight control system in operation, which helps to combat environmental disturbances.
2. It is critical for the drone to land accurately in a fixed position for landed wireless charging. If the landing point deviates too much from the expected position, the charging performance will be greatly affected, and it may be difficult to start the motor again to adjust the position. In the UAV hover wireless charging system, the in-flight drone can dynamically adjust its position relative to the transmitter for the best charging performance.
3. The UAV hover wireless charging system provides a seamless charging experience to prolong the UAV's mission: we can literally charge commercial-grade drones in minutes while in flight. These drones no longer depend on a dedicated landing spot, and we can charge many drones at once if we have enough power at the power source. Although the drone *can* land on a specific pad for charging, considering that landing and take-off are the most risky events (as in any aspect of aviation), and also energy- and time-consuming, it is preferable to keep the drone hovering for few minutes instead.

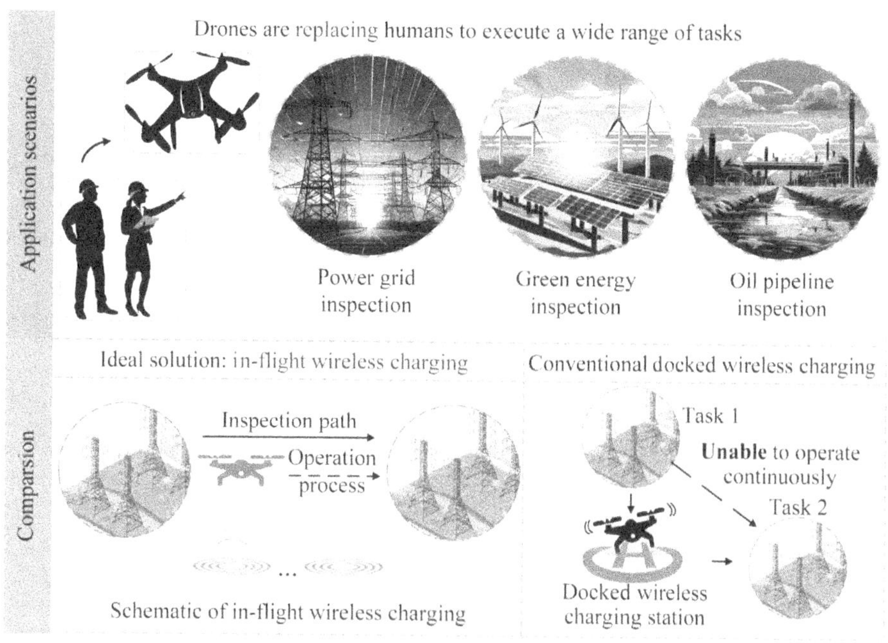

Figure 2.7 Advantages of a UAV hover wireless charging system.

2.3 Challenges of Hovering Wireless Charging

Although hover wireless charging technology can be a suitable solution for UAVs to recharge, it still faces some unique technical challenges for use in industrial applications, which merely transplanting wireless power transmission technology from electric vehicles to UAVs will not solve. Wireless in-flight charging technology provides a flexible replenishment method for UAVs, and has immensely promising application prospect in the fields of power grid and pipeline inspection. However, wireless in-flight charging has three particular technical problems to solve: (1) continuous perturbation of coupling strength, (2) lightweight UAV-side design, and (3) rapid energy replenishment.

1. In inductive power transfer (IPT) systems, the mutual inductance between the coupling coils is affected by their relative positions [20]. When the gap, angle, and transverse and longitudinal positions between the coils change, the coupling strength also changes. This inevitably affects the stability of electrical energy transmission. Unlike the predictable horizontal position in electric vehicle wireless charging, hover wireless charging must continuously and in real time account for vertical, horizontal, and angular deviations. To explore the effect of the environment on drone charging, the DJI Matrice 600 Pro was used to test hovering in a breeze. Figure 2.8 shows the 3D position offset data measured with a sampling time of 20 ms using the ultra-wideband positioning system and presented in a multidimensional view. It can be seen

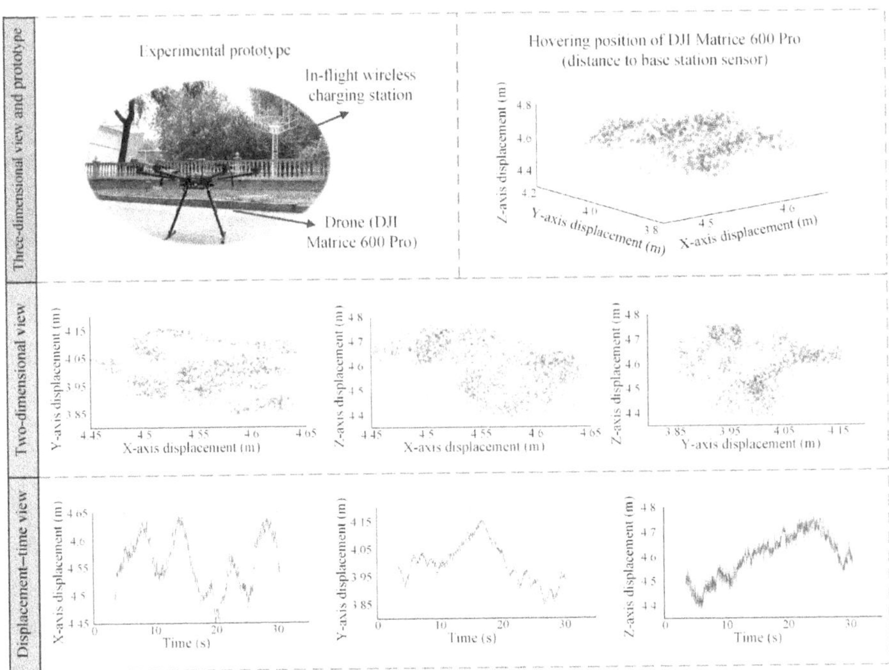

Figure 2.8 Hovering UAV position fluctuation test results.

that the maximum position offsets of the hovering UAV in the x, y, and z directions are about 200 mm, 300 mm, and 400 mm, respectively, over 30 s. Despite the favourable weather conditions during the test, the hovering position of the UAV still shows random fluctuations over a wide area. For hover wireless charging, this must lead to irregular continuous perturbation of the mutual inductance.

Obviously, these particular perturbations do not exist in electric vehicle applications [21], and hence some traditional technical solutions cannot be straightforwardly applied to hover wireless charging systems. In summary, continuous perturbation of coupling strength is the first key technical issue of wireless in-flight charging.

2. As an illustration of the need for lightweight UAV-side design, the DJI Matrice 600 Pro is configured with a TB47S battery with a total unladen UAV weight of 9.5 kg [22]. Figure 2.9 shows the data for the DJI Matrice 600 Pro's endurance hours under different loads based on the TB47S battery configuration [22]. It can be seen that the flight time of the UAV in the unloaded condition is 32 minutes, while in the fully loaded condition it is significantly shorter at 16 minutes. Therefore, when the load of the UAV increases, power loss during hovering also increases, and so reduces the hovering time. In other words, unlike conventional wireless charging systems, for hover wireless charging systems, particular attention must be paid to the additional power consumption caused by any added energy receiver unit weight. Although magnetic materials can be used to enhance the system coupling strength for wireless charging of electric vehicles [23], this solution may not be suitable due to the limited load capacity of UAVs.

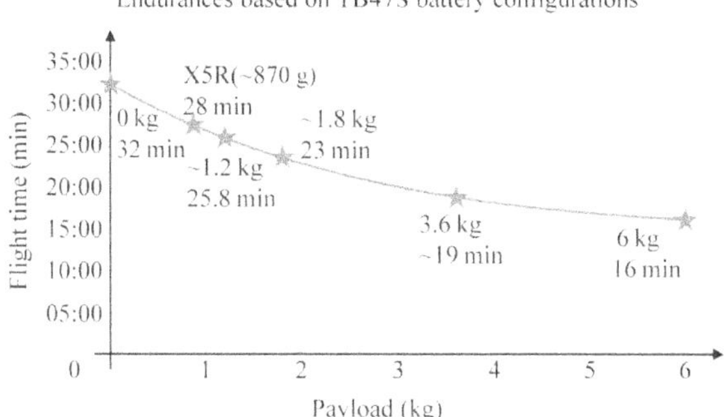

Figure 2.9 The endurance of the DJI Matrice 600 Pro under different payloads [22].

Some detection or control units facilitate the detection of relevant links in the IPT system. However, such solutions are bound to increase the system complexity and weight. For hover charging systems, the weight increase on the UAV side needs to be carefully considered. In summary, a lightweight UAV-side design is the second key technical issue for wireless in-flight charging.

3. UAVs have recently been widely used for daily tasks in a variety of industrial applications with the aim of increasing work efficiency and reducing labour costs. Hovering wireless charging is conducive to effectively extending UAV range. For electric vehicles, wireless charging allows for fast and slow charging [24]. For UAV hover charging, fast charging is essential due to the power loss of the UAV when hovering. The required electrical energy replenishment needs to be accomplished within a short hovering time that cannot exceed the time allowed for UAV hovering operations. This adds new requirements for the design of hover wireless charging systems.

Unlike the standardization and systematization of electric vehicle wireless charging, the wide variability of UAV models and styles makes it impossible to generalize the relevant parameter indices for hovering fast-charging design – that is, it is necessary to carry out targeted design according to the use of different aircraft. To summarize, rapid energy replenishment is the third key technical problem faced by wireless in-flight charging.

In addition to the three major technical hurdles described above, the configuration and implementation of hover wireless charging is also important. UAVs may perform inspection tasks on power grids and other lines in remote locations; to accommodate such scenarios, hover charging systems are recommended to be designed as standa-lone distributed energy systems rather than drawing power from the grid. Therefore, the combination of photovoltaic [25] and hover charging to achieve the integration of photovoltaic storage and charging is another technical challenge.

2.4   Solutions for Special Challenges

In recent research on WPT, researchers have designed solutions for different applications to enhance the performance of wireless charging, including in the areas of misalignment tolerance, transmission distance, power, efficiency, and robustness. As mentioned above, unlike other applications [26–28], the hover wireless charging system of UAVs faces key technical issues and thus it may not be possible to directly adopt solutions from other applications. In this section, we focus on the three key technical issues mentioned above and introduce corresponding solutions for hover wireless charging systems.

2.4.1 Continuous Disturbance of Mutual Inductance

Unmanned aerial vehicles are subject to environmental influences such as high winds when hovering charging, which leads to continuous fluctuations in the relative positions of the coils in the WPT system. In order to ensure the safety and stability of wireless charging, it is necessary to overcome the effects of continuous changes in (positional) mutual inductance.

There are two schemes to overcome the effects of continuous fluctuations in mutual inductance on UAV wireless charging systems: the coil optimal design and the control scheme. The coil topology design will be described in detail in Chapter 3, and the control schemes will be introduced in Chapters 4–6. It is worth noting that control solutions are better suited to overcoming mutual inductive continuous perturbations because:

1. Although the optimized design of the coil can improve the system's misalignment resistance performance to a certain extent, the installation of the coil is difficult due to the limitations of the UAV's installation space and load capacity.
2. The improvement of system performance by coil design optimization is limited [29]. The position of the UAV changes over a wide range when hovering, and it is harder to overcome the large-scale continuous perturbation of mutual inductance by coil design alone; a control scheme is still needed to allow for stable output of the system.

2.4.2 Lightweight Pickup Unit

Due to the limited load capacity of UAVs, lightweight requirements must be met during the design of the receiving side of the WPT system. As mentioned earlier, UAV endurance is strongly affected by the load capacity of the UAV, and the load and space are further limited when the UAV carries additional mission equipment. Therefore, it is necessary to limit the maximum carrying weight for the specific mission of the UAV. The lightweight design of the receiver side of the WPT system is particularly important as any additional weight will increase the power loss and so reduce the UAV's endurance. Taking the DJI Matrice 600 Pro as an example, with six TB47S battery packs the UAV can fly with a Zenmuse X5R camera for 28 minutes [22].

Extra time needs to be reserved to ensure that the UAV performs its tasks safely in grid inspection scenarios. The reserved time consists of the following components:

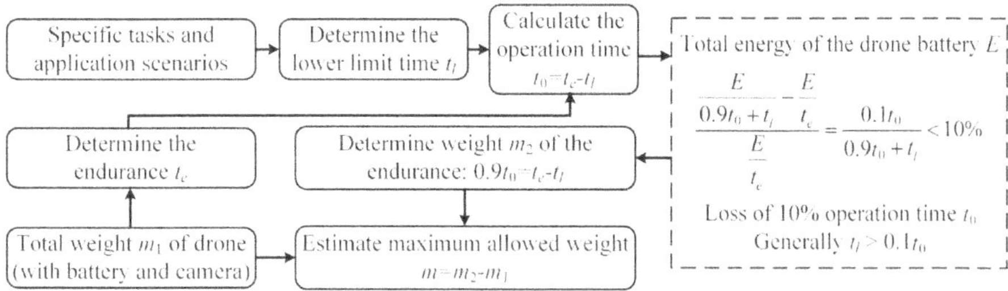

Figure 2.10 Flowchart for calculating the maximum load of the UAV.

round-trip time from the take-off/landing position to the inspection position, plus a safety event reserved at the lower battery limit. Here we design it as 6 minutes, so the actual UAV flight time is 22 minutes. We limit the additional power consumption caused by the WPT receiver device to 10% (i.e. the power consumption caused by the receiver device is less than 10% of the UAV's own power consumption). Therefore, the hovering time of a drone with a WPT receiver device should be greater than 19.8 minutes, yielding a maximum weight of about 330 g for the receiver device allowed for the DJI Matrice 600 Pro. Based on this analysis, a flowchart for estimating the maximum allowable weight of a receiving device for a WPT system is presented in Fig. 2.10. This weight can be estimated accurately depending on the specifics of the mission performed by the UAV.

The upper weight limit of the WPT receiver device can be determined using the approach described above. In order to limit the weight of the receiver side to below the upper weight limit, previous studies have focused on the coil optimal design. Coil optimization requires the consideration of mounting position, shape, and material of the receiver coil for different UAV shapes. In addition to coil optimization, the minimization design of the receiver-side circuit is also an important aspect of a receiver side lightweight design. The transmitter-side control scheme does not require any additional device on the UAV and is more suitable for UAV applications. Transmitter control without communication to achieve constant output is difficult. This method of transmitter control is the focus of this book, and will be introduced in Chapters 4–6. In summary, coil topology optimization design as well as reliable single-side control are two important solutions for the lightweight design of a UAV wireless in-flight charging system while meeting the UAV take-off weight limit.

2.4.3　Fast Charging

In everyday life, people often charge their phones, tablets, and computers while using them, and this can be done with hovering UAVs as well. These two scenarios are similar but different in that the UAV charging system is more challenging as it needs to take into account the energy loss associated with hovering and the time constraints of mission execution. To address these challenges, there are two key questions that need to be answered: how long does it take to hover fast charging? And how do we build a suitable system to realize hover fast charging?

In general, the hovering time of the UAV in a fixed position during power inspection is not very long – only a few minutes – so it needs to complete charging quickly. Using our DJI Matrice 600 Pro data as an example, when loaded with a 330 g energy receiver it has a flight time of about 25.8 minutes on a full charge. It can be estimated that the average power consumption of the UAV is 1400 W based on a total of 600 Wh of energy from six TB47S batteries [22]. The energy needs to be replenished to the maximum extent possible, as permitted by the characteristics of the battery. The maximum charging power of each of the TB47S batteries is 180 W, so the total charging power of the six batteries is 1080 W. The charging power should be considered as the sum of the two parts of the hovering loss and the additional power supplementation. This means the power of the fast-charging system should be 2480 W.

In order to realize an in-flight fast-charging system, the goals of high frequency, high power, and high efficiency need to be met [30]. Compared to silicon-based semiconductors, wide-band semiconductors (i.e. gallium nitride (GaN) and silicon carbide (SiC)) have significant advantages in terms of their higher operating temperatures, higher voltages, higher frequencies, and lower parasitic parameters [31]. Undoubtedly, power switching tubes composed of wide-bandwidth devices are especially suitable for fast-charging systems [32]. Silicon carbide is more advantageous in high-voltage, high-power scenarios, and GaN is more suitable for high-frequency applications. For a UAV grid inspection system, a kilowatt-level and several hundred kilohertz WPT system is generally required, so GaN is preferred.

In summary, there are two main parts to realizing fast in-flight wireless charging for electrically powered UAVs. One part is to determine the charging power based on the charging demand and hovering loss of the UAV; the other part is to design and develop a suitable power converter based on the frequency and power requirements of the WPT system and the application scenarios.

2.5   System Construction for Hovering Wireless Charging

Unmanned aerial vehicles are usually used for larger working area ranges, such as remote areas. These scenarios can mean we need to deal with the problem of a stable energy supply. For grid inspection application scenarios, a Rogowski coil can be designed to provide an electrical energy supply [33]. Similar to a current transformer, it can easily capture electrical energy from the induced magnetic field generated by transmission lines. For other application scenarios, such as oil or gas pipeline inspection, there is no suitable source for power acquisition. Therefore, to meet the demand for UAV hover charging, it is necessary to develop photovoltaic integrated energy storage and charging (PIESC) technology, a standalone modular power source with a promising future. The design idea of the UAV wireless charging station based on PIESC technology is shown in Fig. 2.11. The system should operate in off-grid mode

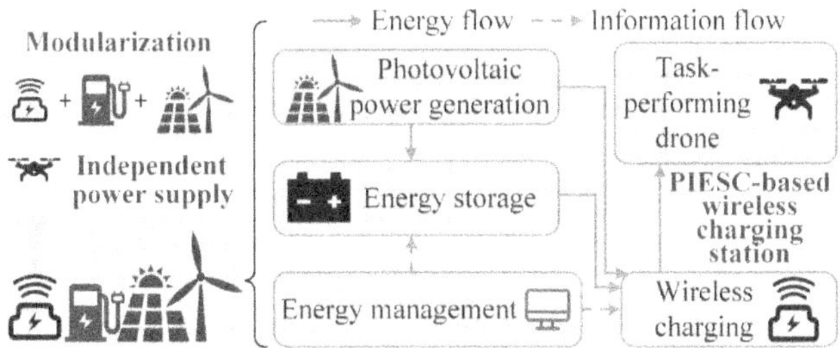

Figure 2.11 Schematics of a UAV wireless charging station based on PIESC technology.

and act as an individual system that is widely distributed and covers the various work areas of UAV inspection.

As depicted in Fig. 2.11, the design and location of the charging station needs to take into account the inspection routes of the UAV, the endurance time, and the hovering time at the inspection task points. The PIESC charging stations are recommended to be developed at work points along the inspection routes that require longer inspection times, and the number of stations to be developed needs to be decided by combining the cost of construction and the cost of daily maintenance. To summarize, PIESC-based wireless charging technology can greatly enhance the practicality and universality of in-flight wireless charging of UAVs.

2.6   Summary

The related development of artificial intelligence and other technologies has greatly facilitated the wide application of UAVs in various industries. Wireless in-flight charging shows great promise for overcoming the challenge of short endurance times of UAVs. This chapter summarizes the special challenges that make this system different from other WPT applications, which consist of the continuous variation of coupling coefficients, lightweight UAV-side design, and wireless fast charging. The solutions to these three key issues are further explored and the PIESC-based charging station for UAVs is presented.

References

[1] Fact.MR. Drone market outlook (2024 to 2034). Available: www.factmr.com/report/62/drone-market

[2] S. H. Chung, B. Sah, and J. Lee, 'Optimization for drone and drone–truck combined operations: a review of the state of the art and future directions', *Comput. Oper. Res.*, vol. 123, pp. 1–29, Jun. 2020.

[3] Intelligence Research Group. Intelligence research consulting. Available: www.chyxx.com

[4] H. Huang and A. V. Savkin, 'A method of optimized deployment of charging stations for drone delivery', *IEEE Trans. Transp. Electrific.*, vol. 6, no. 2, pp. 510–518, Jun. 2020.

[5] N. Ahmadian, G. J. Lim, M. Torabbeigi, and S. J. Kim, 'Smart border patrol using drones and wireless charging system under budget limitation', *Comput. Ind. Eng.*, vol. 164, Art. no. 107891, Feb. 2022.

[6] S. Assawaworrarit, X. Yu, and S. Fan, 'Robust wireless power transfer using a nonlinear parity-time-symmetric circuit', *Nature*, vol. 546, no. 7658, pp. 387–390, Jun. 2017.

[7] R. Mai, Y. Liu, Y. Li, P. Yue, G. Cao, and Z. He, 'An active-rectifier-based maximum efficiency tracking method using an additional measurement coil for wireless power transfer', *IEEE Trans. Power Electron.*, vol. 33, no. 1, pp. 716–728, Jan. 2018.

[8] M. Liu, H. Zhang, Y. Shao, J. Song, and C. Ma, 'High-performance megahertz wireless power transfer: topologies, modeling, and design', *IEEE Ind. Electron. Mag.*, vol. 15, no. 1, pp. 28–42, Mar. 2021.

[9] S. Cui, Z. Wang, S. Han, C. Zhu, and C. C. Chan, 'Analysis and design of multiphase receiver with reduction of output fluctuation for EV dynamic wireless charging system', *IEEE Trans. Power Electron.*, vol. 34, no. 5, pp. 4112–4124, May 2019.

[10] Z. Zhang, W. Ai, Z. Liang, and J. Wang, 'Topology-reconfigurable capacitor matrix for encrypted dynamic wireless charging of electric vehicles', *IEEE Trans. Veh. Technol.*, vol. 67, no. 10, pp. 9284–9293, Oct. 2018.

[11] S. Y. Hui, 'Planar wireless charging technology for portable electronic products and Qi', *Proc. IEEE*, vol. 101, no. 6, pp. 1290–1301, Jun. 2013.

[12] Z. Zhang, H. Pang, A. Georgiadis, and C. Cecati, 'Wireless power transfer: an overview', *IEEE Trans. Ind. Electron.*, vol. 66, no. 2, pp. 1044–1058, Feb. 2019.

[13] I. A. Mashhadi, M. Pahlevani, S. Hor, H. Pahlevani, and E. Adib, 'A new wireless power-transfer circuit for retinal prosthesis', *IEEE Trans. Power Electron.*, vol. 34, no. 7, pp. 6425–6439, Jul. 2019.

[14] Y. Gu, J. Wang, Z. Liang, Y. Wu, C. Cecati, and Z. Zhang, 'Single-transmitter multiple-pickup wireless power transfer: advantages, challenges, and corresponding technical solutions', *IEEE Ind. Electron. Mag.*, vol. 14, no. 4, pp. 123–135, Dec. 2020.

[15] C. Cai, J. Wang, F. Zhang, X. Liu, P. Zhang, and Y. G. Zhou, 'A multichannel wire-less UAV charging system with compact receivers for improving transmission stability and capacity', *IEEE Syst. J.*, vol. 16, no. 1, pp. 997–1008, Mar. 2021.

[16] Y. Gu, J. Wang, Z. Liang, and Z. Zhang, 'Communication-free power control algorithm for drone wireless in-flight charging under dual-disturbance of mutual inductance and load', *IEEE Trans. Ind. Inform.*, vol. 20, no. 3, pp. 3703–3714, Mar. 2024.

[17] Y. Gu, J. Wang, Z. Liang, and Z. Zhang, 'Mutual-inductance-dynamic-predicted constant current control of LCC-P compensation network for drone wireless in-flight charging', *IEEE Trans. Ind. Electron.*, vol. 69, no. 12, pp. 12710–12719, Dec. 2022.

[18] Y. Gu, J. Wang, Z. Liang, and Z. Zhang, 'A wireless in-flight charging range extended PT-WPT system using S/single-inductor-double-capacitor compensation network for drones', *IEEE Trans. Power Electron.*, vol. 38, no. 10, pp. 11847–11858, Oct. 2023.

[19] T. Campi, S. Cruciani, F. Maradei, and M. Feliziani, 'Innovative design of drone landing gear used as a receiving coil in wireless charging application', *Energies*, vol. 12, no. 18, p. 3483, 2019.

[20] S. Babic, F. Sirois, C. Akyel, and C. Girardi, 'Mutual inductance calculation between circular filaments arbitrarily positioned in space: alternative to Grover's formula', *IEEE Trans. Magn.*, vol. 46, no. 9, pp. 3591–3600, Sep. 2010.

[21] X. Dai, J. Jiang, and J. Wu, 'Charging area determining and power enhancement method for multiexcitation unit configuration of wirelessly dynamic charging EV system', *IEEE Trans. Ind. Electron.*, vol. 66, no. 5, pp. 4086–4096, May 2019.

[22] DJI. Support for Matrice 600 Pro. Available: www.dji.com/matrice600-pro

[23] S. Li and C. C. Mi, 'Wireless power transfer for electric vehicle applications', *IEEE J. Emerg. Sel. Topics Power Electron.*, vol. 3, no. 1, pp. 4–17, Mar. 2015.

[24] S. Rivera, S. Kouro, S. Vazquez, S. M. Goetz, R. Lizana, and E. Romero-Cadaval, 'Electric vehicle charging infrastructure: from grid to battery', *IEEE Ind. Electron. Mag.*, vol. 15, no. 2, pp. 37–51, Jun. 2021.

[25] M. ElSayed, A. Foda, and M. Mohamed, 'Autonomous drone charging station planning through solar energy harnessing for zero-emission operations', *Sustain. Cities Soc.*, vol. 86, Art. no. 104122, Nov. 2022.

[26] S. Lee and S.-H. Lee, 'dq synchronous reference frame model of a series-series tuned inductive power transfer system', *IEEE Trans. Ind. Electron.*, vol. 67, no. 12, pp. 10325–10334, Dec. 2020.

[27] J. P.-W. Chow, H. S.-H. Chung, and C.-S. Cheng, 'Use of transmitter-side electrical information to estimate mutual inductance and regulate receiver-side power in wireless inductive link', *IEEE Trans. Power Electron.*, vol. 31, no. 9, pp. 6079–6091, Sep. 2016.

[28] L. Zhao, D. J. Thrimawithana, U. K. Madawala, A P. Hu, and C. C Mi, 'A misalignment-tolerant series-hybrid wireless EV charging system with integrated magnetics', *IEEE Trans. Power Electron.*, vol. 34, no. 2, pp. 1276–1285, Apr. 2019.

[29] Z.-J. Liao, F. Wu, C.-H. Jiang, Z.-R. Chen, and C.-Y. Xia, 'Analysis and design of ideal transformer-like magnetic coupling wireless power transfer systems', *IEEE Trans. Power Electron.*, vol. 37, no. 12, pp. 15728–15739, Dec. 2022.

[30] K. Chen and Z. Zhang, 'In-flight wireless charging: a promising application-oriented charging technique for drones', *IEEE Ind. Electron. Mag.*, vol. 18, no. 1, pp. 6–16, Mar. 2024.

[31] J. Millán, P. Godignon, X. Perpiñà, A. Pérez-Tomás, and J. Rebollo, 'A survey of wide bandgap power semiconductor devices', *IEEE Trans. Power Electron.*, vol. 29, no. 5, pp. 2155–2163, May 2014.

[32] S. Li, S. Lu, and C. C. Mi, 'Revolution of electric vehicle charging technologies accelerated by wide bandgap devices', *Proc. IEEE*, vol. 109, no. 6, pp. 985–1003, Jun. 2021.

[33] L. Du, C. Wang, X. Li, L. Yang, Y. Mi, and C. Sun, 'A novel power supply of online monitoring systems for power transmission lines', *IEEE Trans. Ind. Electron.*, vol. 57, no. 8, pp. 2889–2895, Aug. 2010.

3 Magnetic Coupler Design for Wireless Charging of UAV

3.1 Introduction

Unmanned aerial vehicles (UAVs) are becoming increasingly popular in everyday life and in military applications. They are used for inspection, transportation, search and rescue, and other tasks [1, 2]. For electrically driven UAVs, the capacity limitations of lithium batteries prevent the UAV from cruising for a long period of time or over a wide range. The conventional practice is to let the UAV fly back to the base for manual battery replacement or wired charging, which greatly affects the UAV's operational efficiency. In recent years, wireless power transfer (WPT) technology has been extensively researched and applied due to its advantages in terms of flexibility, convenience, and efficiency. Applying WPT to UAVs can improve their endurance; methods of WPT are generally categorized into two types: in-flight charging [3–7] and charging after landing [8–11].

In contrast to other applications of WPT [12–19], wireless charging for UAVs faces some special challenges. It is important to consider the special structure of UAVs and their limitations in terms of loads when installing an energy-receiving device, which increases the requirements of the magnetic coupling design of the WPT system. The receiver-side device (including the coupling coil, compensation topology, and control unit) needs to be as lightweight as possible to minimize energy loss when the UAV is hovering, and thus prolong the UAV's endurance.

Another technical challenge for wireless charging systems for UAVs is the problem of frequent misalignment between coils. Since the UAV may hover in the air or land on the charging pad for charging, it faces the problem of possible misalignment between the coupling coils in three aspects: lateral, longitudinal, and angular. Misalignment affects the output of the system and reduces the charging efficiency.

In response to these limitations of UAV payload and space for the magnetic coupling device, as well as the wide range of misalignment problems in the UAV charging system, current research is mainly divided into two aspects: the optimal design of the magnetic coupling structure and a control strategy to prevent misalignment. Research to optimize the design of the UAV magnetic coupling structure has explored some novel design methods of the receiving coil [20–24] and the overall magnetic coupling structure [25–28] to improve the alignment capability of the wireless charging system. This chapter focuses on the coil structures suitable for

UAV wireless charging applications. Various control strategies for constant charging output of UAVs are presented in the next chapter.

3.2 Pickup Coil Design for the UAV Wireless Charging System

3.2.1 Embedded Lightweight Squirrel-Cage Coil

As mentioned earlier, the wireless charging system applied to UAVs needs to be compact and lightweight on the receiving side to ensure it is suitable for the mechanical structure of the UAV and has a certain degree of misalignment resistance. This section introduces a novel squirrel-cage-type receiving coil [20] that can be embedded in the landing gear of a UAV (Fig. 3.1). The proposed embedded coil structure includes a lightweight receiving coil and greatly reduces the wind resistance suffered by the UAV in flight by not mounting on the UAV fuselage. This compact receiving coil structure does not drastically change the structure of the UAV itself and maintains good flight performance, making it appropriate for a UAV wireless charging system.

Figure 3.1 shows the installation of a WPT system suitable for the shape and structure of the UAV and which also satisfies the requirements of high misalignment performance and low weight of the receiving side. In the coupling structure, the receiving-side squirrel-cage coils are mounted at the lower end of each of the two T-structure landing gears. The transmitting coils consist of two coils, internal and external, connected in series. In addition to a large external transmitting coil, a small internal transmitting coil enhances the magnetic coupling strength in the centre region, which ensures high coupling strength of the system at all positions. The current directions of the two transmitting coils should be consistent to enhance the mutual

Figure 3.1 The proposed UAV wireless charging coupler construction.

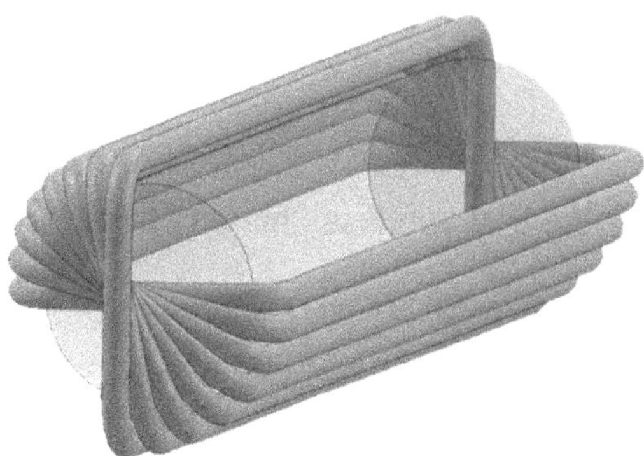

Figure 3.2 Novel squirrel-cage receiver coil.

inductance. The proposed coupling structure is particularly suitable for medium-sized UAV structures.

Figure 3.2 illustrates the proposed squirrel-cage coil on the receiver side. This coil can fully utilize the space of the landing gear and be closer to the transmitter coil on the bottom, ensuring a lightweight and compact design on the UAV side. This structure can receive magnetic flux from horizontal and vertical directions, which to some extent reduces the effect of misalignment on output performance. Using the proposed lightweight and compact coupler structure, the receiver coil can stably receive the magnetic field produced by the transmitter coil and so provide stable and efficient charging performance for the UAV.

The specific data on the coupling mechanism are as follows:

- The outer diameter of the large transmitting coil is 440 mm, with 15 turns.
- The outer diameter of the small transmitting coil is 100 mm, with 5 turns.
- The squirrel-cage receiving coil is wound on a hollow plastic tube of 200 mm diameter and 20 mm length, with 11 turns along the tube, where the first and last turns are perpendicular to each other in the plane.
- The transmitting and receiving coils are wound with 3 mm and 1.5 mm diameter Litz wire, respectively.

The parameters of the coupling mechanism need to be adjusted for different types of UAVs to ensure excellent performance of the charging system.

It has been experimentally verified that the proposed squirrel-cage coil can better cope with misalignment produced by the transmitter coil in the x or z directions compared with a conventional planar helical coil. When the pickup coil is mounted on the UAV landing gear, the spacing between coils in the coupling structure is smaller, which enhances the system's coupling strength. The proposed structure ensures low electromagnetic radiation of the electrical equipment on the UAV side, as well as allowing a lightweight and compact system that does not increase the UAV's air resistance during flight. It is experimentally verified that the system can satisfy a misalignment tolerance of 100 mm, making it appropriate for a UAV wireless charging system.

3.2.2 Hollow Pickup Coil for In-flight UAV

In the previous section, a novel squirrel-cage pickup coil was introduced, which can be used for landed wireless charging of medium-sized UAVs. In this section, a hollow pickup coil is presented [21], which can be employed for hovering wireless charging of small UAVs. As mentioned earlier, in UAV charging systems the UAV-side design needs to fulfil the lightweight requirement. In addition, to ensure high coupling strength, the influence of coil size and shape on the electromagnetic coupling performance of the system needs to be considered. The transmitter coil in the magnetic coupling mechanism has a little more freedom and does not have such a severe weight limitation. In summary, the lightweight and compact construction of the receiving coil is especially important for a hovering wireless charging system, and the coil size and shape need to be optimized to reduce the influence of the coil on the aerodynamic behaviour of the UAV.

Fig. 3.3 depicts a lightweight design solution for a megahertz wireless charging system in which the receiving coil is built within a copper-plated plastic structure. The receiving coil mounted on the UAV is close to the propeller shroud of the UAV. Unlike the coil structure in conventional WPT systems, the proposed receiving coil avoids the use of ferromagnetic materials, which reduces its impact on the aerodynamic performance of the UAV. A high coupling coefficient ($k > 0.05$) of the WPT system is ensured by the optimized design of the coupling coil. Therefore, the proposed coupling mechanism ensures a lightweight and highly efficient wireless charging system while meeting the aerodynamic performance requirements of the UAV.

In the coupling mechanism shown in Fig. 3.3, the transmitting coil has an outer diameter of 20 cm wrapped around two turns; it uses a printed circuit board type of coil that meets the needs of a megahertz WPT system. For the coupling mechanism, a closed-loop control algorithm is designed in the system circuit to resist dynamic changes in load and mutual inductance, and to maintain a stable system output. Experiments have proven that the magnetic field generated by the system does not affect the flight control of the UAV and does not cause an increase in the temperature

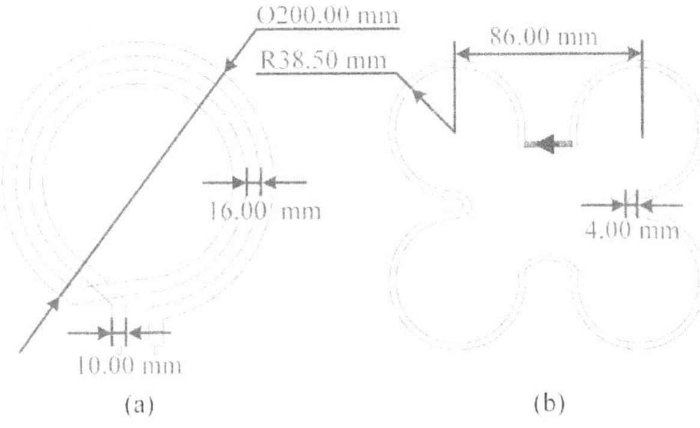

Figure 3.3 System coils: (a) transmitter; (b) receiver.

of the UAV fuselage. Therefore, the proposed wireless charging scheme can provide a safe and stable energy supply for the UAV without negative effect on its flight capabilities.

3.2.3   Onboard Integration-Based Pickup Coil

A wireless charging system for UAVs faces the challenge of UAV attitude changes – as well as positional misalignments – thus requiring high misalignment tolerance in the WPT system to account for possible perturbation. In this section we present a novel magnetically integrated WPT system suitable for UAVs. This system is capable of achieving high misalignment tolerance and constant current output [22]. Figure 3.4 illustrates the magnetic coupling structure of the proposed WPT system, with double-sided LCC high-order topology. The design scheme of the transmitter coil, receiver coil, and integrated coil improve the system's misalignment tolerance to ensure a stable output current can be obtained when there is misalignment. The proposed magnetic coupling structure also offers a lightweight and compact UAV-side receiving device.

Two identical receiving coils are used in this integrated coil structure, with the compensation coil integrated near the receiving coils. No ferromagnetic material is used, which ensures the coupling strength of the WPT system and reduces the size and weight of the receiving device. For a landed UAV wireless charging system there may be some random deviations in the landing position of the UAV, so the system misalignment tolerance capability needs to be good to ensure stable energy transmission. In the proposed coupling mechanism, both the transmitting and receiving coils are circular to avoid the influence of angular deviation on transmission performance.

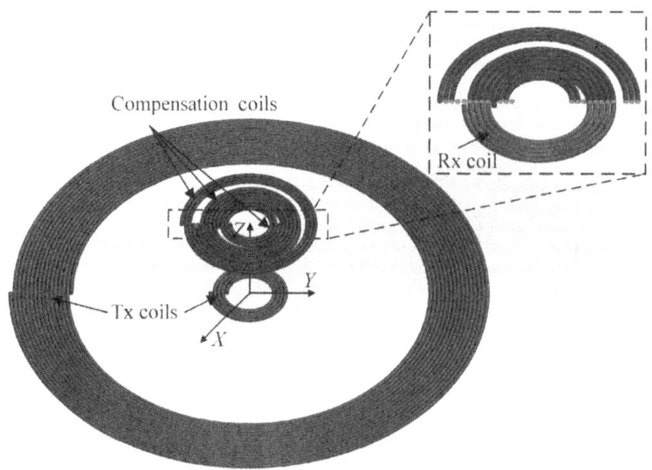

Figure 3.4 The proposed coil model for the UAV.

The problem of electromagnetic radiation in WPT systems is an important challenge, and traditional electromagnetic shielding schemes mainly use a combination of aluminium plates and ferrites to suppress magnetic field leakage from WPT systems. However, the eddy current loss generated by the alternating magnetic field on the shielding mechanism in this scheme reduces the system efficiency, while the weight of the ferrite and aluminium plate aggravates the energy loss of the UAV. To avoid these problems, flexible nanocrystalline core materials can be used in the coupling mechanism [24]. Therefore, the proposed electromagnetic shielding scheme employs a three-layer shielding structure of nanocrystals, ferrite, and aluminium foil, which is applied to the receiving coil to prevent the electronic devices on the UAV receiving interference from the wireless charging system.

It can be demonstrated that the proposed magnetic coupling mechanism achieves stable output power and efficiency under a lateral misalignment of 150 mm, and the constant output under load variations can be satisfied, which also meets the lightweight requirements of the UAV-side receiver device.

3.3 Magnetic Coupling Structure for the UAV Wireless Charging System

3.3.1 Orthogonal Magnetic Coupler Structure

In UAV wireless charging systems, compact and lightweight pickup coils are usually used to reduce the UAV-side weight while optimizing the magnetic field distribution. As shown in Fig. 3.5, a new orthogonal magnetic coupling structure is applied to the UAV wireless charging system [25, 26]. The transmitting coil includes two rectangular coils with a flat U-shaped ferrite core and a layer of aluminium plate. The two transmitting coils are wrapped around the two side pillars of the ferrite core to ensure the high coupling strength of the WPT system, while the aluminium plate can

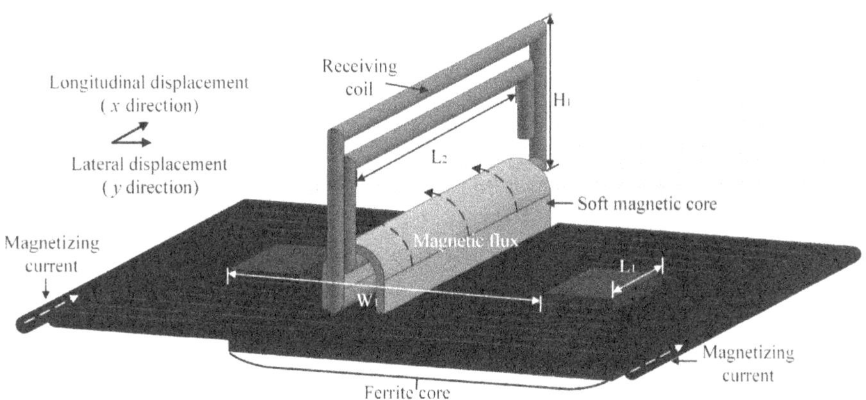

Figure 3.5 The proposed orthogonal magnetic structure.

effectively achieve electromagnetic shielding to prevent the WPT system affecting the UAV's electronics.

Significantly, the receiving coil includes a rectangular coil and a flexible nanocrystalline material that is tightly wound and vertically placed along the UAV landing gear. This topology means that, since the UAV landing gear is generally vertical, the proposed receiver coil structure can be embedded in the landing gear without increasing the UAV's wind resistance during flight. In addition, the proposed coil structure ensures the system has high coupling strength, hence improving stability and efficiency.

Adding ferromagnetic materials such as ferrite is generally a viable option to increase system magnetic coupling strength. However, for UAV wireless charging systems, ferrite may not be suitable due to its weight and not being soft. In the proposed magnetic coupling mechanism, a soft and lightweight iron-based nanocrystalline magnetic material is used to enhance the coupling strength of the system. The iron-based nanocrystalline material is generally cut, thermoformed, and multilayer-laminated to obtain suitable specification parameters. Compared with PC95 ferrite material, this nanocrystalline magnetic material has lower core loss and better flexibility, making it more suitable for UAV wireless charging systems. When wrapped around the lower edge of the receiving coil it can enhance the system coupling strength and thus improve the transfer behaviour of the wireless charging system.

The proposed orthogonal-type coupling coil structure with flexible nanocrystals can achieve high transmission efficiency and electromagnetic radiation in WPT systems. In addition, a wireless charging platform that enables position correction assistance has been developed to ensure the UAV can land accurately on the wireless charging pad [25]. Significantly, the system does not use a coil position detection device to realize the accurate landing of the UAV. When the UAV lands on the charging pad, contactless energy replenishment begins. The proposed system architecture is utilized to ensure flexible, stable wireless charging and to enhance the range of the UAV.

3.3.2 Free-Rotation Asymmetric Magnetic Coupling Structure

Although the orthogonal magnetic structure described in the previous section allows for a lightweight coil structure with low magnetic field leakage, it still requires position correction aids to achieve a more stable output power, and thus the structure has limited misalignment tolerance. In this section, a novel asymmetric magnetic coupler for conformal lightweight UAV-side pickup structures is proposed to simultaneously address the rotational and horizontal misalignment challenges of UAVs [27]. In the proposed transmitter coil structure, multiple sets of circular coils are connected in series to produce a uniform horizontal magnetic field. The receiving coil is designed as a solenoid valve structure wound along the UAV landing gear; it is covered with a layer of flexible nanocrystalline material on the outer side. The proposed coupled coil structure ensures higher misalignment tolerance of the UAV wireless charging system in horizontal and rotational directions.

Figure 3.6 illustrates the proposed magnetic coupling structure. The pickup coil is wrapped around the UAV's landing gear to enhance the system's coupling strength.

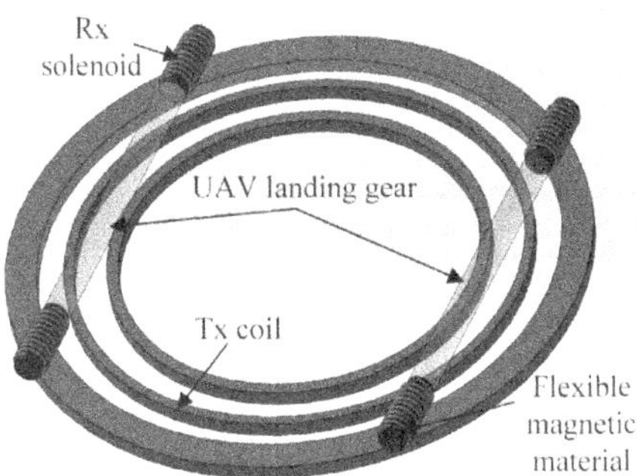

Figure 3.6 The magnetic coupling structure of the proposed system.

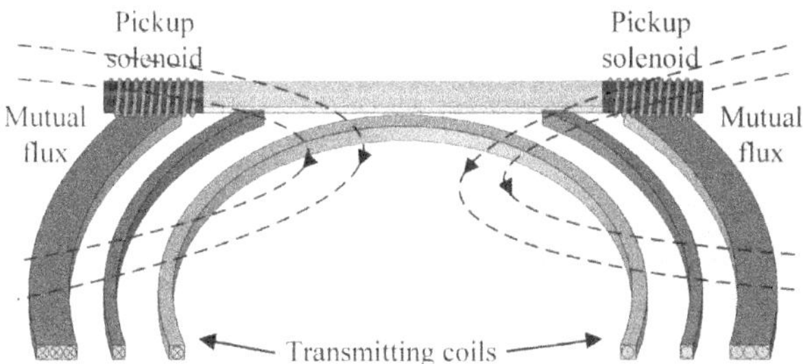

Figure 3.7 A schematic of the horizontal magnetic field.

The helical-tube-type receiving coil can expand the receiving area of the magnetic field. It can be seen in the figure that multiple transmitter coils are connected in series to ensure stable output characteristics for special cases, such as when the UAV landing position is misaligned. The four receiving coils are wound into a solenoid along the UAV landing gear and connected in series, while the flexible nanocrystalline material is wrapped around the inside of the receiving coils to enhance the system's coupling strength.

Figure 3.7 depicts the structure of the system and the magnetic vector diagrams. To avoid the effects of changes in the position of the receiver coils, two receiving coils on the same landing gear are wound in opposite directions. The detailed design optimization scheme of the magnetic coupling mechanism is presented in [27]. A 200 W WPT system experimental platform has been constructed to verify this magnetic coupling structure, and has demonstrated that it achieves stable output under a lateral misalignment of 20 mm, as well as self-rotation.

3.3.3 Compact Omnidirectional Magnetic Structure

The compactness of the receiver unit on the UAV side and the system's misalignment tolerance are two important challenges for UAV WPT systems. A novel magnetic coupling structure is proposed in [28], in which four toroidal coils form the transmitter

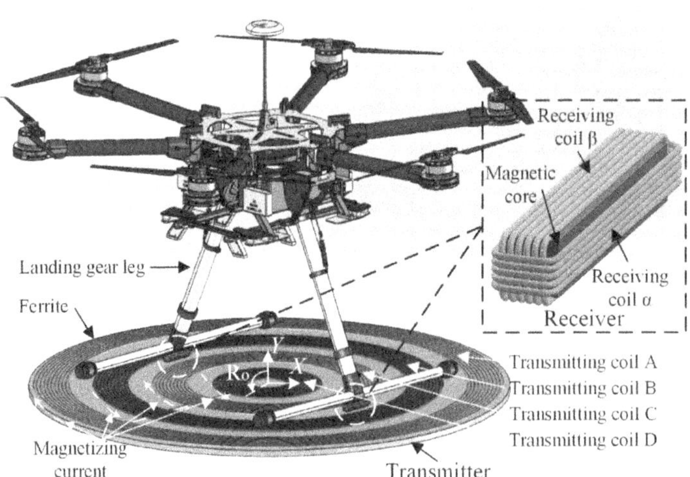

Figure 3.8 General overview of the proposed magnetic structure.

coil. The transmitter coil, magnetized in bipolar mode, ensures the omnidirectional distribution of the magnetic field (i.e. the parallel and perpendicular magnetic fluxes complement each other). On the UAV side, two orthogonal-type receiving coils are integrated to maintain a strong output.

Fig. 3.8 depicts the proposed magnetic coupling structure, in which four toroidal coils with different diameters form a transmitting coil similar to a 'target'. A ferromagnetic material is mounted on the back of the coils to enhance the system's coupling strength. In this WPT system, four transmitter coils are interconnected in series, where A and C are twisted clockwise and B and D are twisted counter-clockwise, which realizes the generation of parallel and perpendicular fluxes and enhances the system's omnidirectional charging capability. The number of transmitter coils can be optimally designed to ensure high misalignment tolerance depending on the practical application.

Two identical receivers are mounted on each wing of the UAV landing gear, where each receiver includes a lightweight and flexible nanocrystalline rectangular core as well as an orthogonal-type receiver coil wound on the outside, which ensures high coupling strength of the WPT system. The perpendicular orthogonal-type receiving coils can receive parallel and perpendicular magnetic fluxes from the transmitting side. The advantages of the proposed coupling mechanism are:

1. The receiver unit is small and lightweight, embedded in the landing gear of the UAV. It can fulfil medium to high power requirements.
2. Only one inverter is required on the transmitter side, eliminating the need for complex control schemes.
3. The designed magnetic coupling structure does not need to be mechanically fixed and has a high misalignment tolerance to enhance the system's robustness.

4. Since the receiving coil is mounted on the landing gear, it ensures high coupling strength while being far away from the UAV's electronic equipment, ensuring good electromagnetic compatibility.

References

[1] Z. Zhang, H. Pang, A. Georgiadis, and C. Cecati, 'Wireless power transfer: an overview', *IEEE Trans. Ind. Electron.*, vol. 66, no. 2, pp. 1044–1058, Feb. 2019.

[2] Y. Gu, J. Chen, S. Chang, and Z. Zhang, 'Constant power control against M/R with expanded PT-symmetric range for wireless in-flight charging of drones', *IEEE Trans. Magn.*, 2023, doi: 10.1109/TMAG.2023.3284826.

[3] W. Han, K. T. Chau, C. Jiang, W. Liu, and W. H. Lam, 'Design and analysis of quasi-omnidirectional dynamic wireless power transfer for fly-and-charge', *IEEE Trans. Magn.*, vol. 55, no. 7, pp. 1–9, Jul. 2019.

[4] Y. Gu, J. Wang, Z. Liang, and Z. Zhang, 'A wireless in-flight charging range extended PT-WPT system using S/single-inductor-double-capacitor compensation network for drones', *IEEE Trans. Power Electron.*, vol. 38, no. 10, pp. 11847–11858, Oct. 2023.

[5] K. Chen and Z. Zhang, 'In-flight wireless charging: a promising application-oriented charging technique for drones', *IEEE Ind. Electron. Mag.*, vol. 18, no. 1, pp. 6–16, Mar. 2024.

[6] Y. Gu, J. Wang, Z. Liang, and Z. Zhang, 'Mutual-inductance-dynamic-predicted constant current control of LCC-P compensation network for drone wireless in-flight charging', *IEEE Trans. Ind. Electron.*, vol. 69, no. 12, pp. 12710–12719, Dec. 2022.

[7] Y. Gu, J. Wang, Z. Liang, and Z. Zhang, 'Communication-free power control algorithm for drone wireless in-flight charging under dual-disturbance of mutual inductance and load', *IEEE Trans. Ind. Inform.*, vol. 20, no. 3, pp. 3703–3714, Mar. 2024.

[8] C. Cai, J. Wang, F. Zhang, X. Liu, P. Zhang, and Y. G. Zhou, 'A multichannel wireless UAV charging system with compact receivers for improving transmission stability and capacity', *IEEE Syst. J.*, vol. 16, no. 1, pp. 997–1008, Mar. 2021.

[9] Y. Li, W. Sun, J. Liu, Y. Liu, X. Yang, and Y. Li, 'A new magnetic coupler with high rotational misalignment tolerance for unmanned aerial vehicles wireless charging', *IEEE Trans. Power Electron.*, vol. 37, no. 11, pp. 12986–12991, Nov. 2022.

[10] C. Rong, X. He, Y. Wu, Y. Qi, R. Wang, and Y. Sun, 'Optimization design of resonance coils with high misalignment tolerance for drone wireless charging based on genetic algorithm', *IEEE Trans. Ind. Appl.*, vol. 58, no. 1, pp. 1242–1253, Jan. 2022.

[11] H. Zhang, Y. Chen, C. H. Jo, S. J. Park, and D. H. Kim, 'DC-link and switched capacitor control for varying coupling conditions in inductive power transfer system for unmanned aerial vehicles', *IEEE Trans. Power Electron.*, vol. 36, no. 5, pp. 5108–5120, May 2021.

[12] Y. Gu, J. Wang, Z. Liang, Y. Wu, C. Cecati, and Z. Zhang, 'Single-transmitter multiple-pickup wireless power transfer: advantages, challenges, and corresponding technical solutions', *IEEE Ind. Electron. Mag.*, vol. 14, no. 4, pp. 123–135, Dec. 2020.

[13] C. Liu, C. Jiang, J. Song, and K. T. Chau, 'An effective sandwiched wireless power transfer system for charging implantable cardiac pacemaker', *IEEE Trans. Ind. Electron.*, vol. 66, no. 5, pp. 4108–4117, May 2019.

[14] X. Dai, J. Jiang, and J. Wu, 'Charging area determining and power enhancement method for multiexcitation unit configuration of wirelessly dynamic charging EV system', *IEEE Trans. Ind. Electron.*, vol. 66, no. 5, pp. 4086–4096, May 2019.

[15] P. Zhang, M. Saeedifard, O. C. Onar, Q. Yang, and C. Cai, 'A field enhancement integration design featuring misalignment tolerance for wireless EV charging using LCL topology', *IEEE Trans. Power Electron.*, vol. 36, no. 4, pp. 3852–3867, Apr. 2021.

[16] C. Jiang, K. T. Chau, C. H. T. Lee, W. Han, W. Liu, and W. H. Lam, 'A wireless servo motor drive with bidirectional motion capability', *IEEE Trans. Power Electron.*, vol. 34, no. 12, pp. 12001–12010, Dec. 2019.

[17] W. Han, K. T. Chau, Z. Hua, and H. Pang, 'An integrated wireless motor system using laminated magnetic coupler and commutative-resonant control', *IEEE Trans. Ind. Electron.*, vol. 69, no. 5, pp. 4342–4352, May 2022.

[18] X. Qu, W. Zhang, S.-C Wong, and C. K. Tse, 'Design of a current-source-output inductive power transfer LED lighting system', *IEEE J. Emerg. Sel. Topics Power Electron.*, vol. 3, no. 1, pp. 306–314, Mar. 2015.

[19] W. Liu, K. T. Chau, C. H. T. Lee, C. Jiang, W. Han, and W. H. Lam, 'A wireless dimmable lighting system using variable-power variable-frequency control', *IEEE Trans. Ind. Electron.*, vol. 67, no. 10, pp. 8392–8404, Oct. 2020.

[20] P. Cao, Y. Lu, H. Zhang, J. Li, W. Chai, C. Cai, and S. Wu, 'Embedded lightweight squirrel-cage receiver coil for drone misalignment-tolerant wireless charging', *IEEE Trans. Power Electron.*, vol. 38, no. 3, pp. 2884–2888, Mar. 2023.

[21] J. M. Arteaga, S. Aldhaher, G. Kkelis, C. Kwan, D. C. Yates, and P. D. Mitcheson, 'Dynamic capabilities of multi-MHz inductive power transfer systems demonstrated with batteryless drones', *IEEE Trans. Power Electron.*, vol. 34, no. 6, pp. 5093–5104, Jun. 2019.

[22] J. Wang, R. Chen, C. Cai, J. Zhang, and C. Wang, 'An onboard magnetic integration-based WPT system for UAV misalignment-tolerant charging with constant current output', *IEEE Trans. Transport. Electrific.*, vol. 9, no. 1, pp. 1973–1984, Mar. 2023.

[23] S. Pang, J. Xu, Z. Xie, J. Lu, H. Li, and X. Li, 'Lightweight UAV's wireless power transfer system for constant current charging without secondary feedback control', *IEEE Trans. Veh. Technol.*, vol. 72, no. 12, pp. 15611–15621, Dec. 2023.

[24] M. Xiong, X. Wei, Y. Huang, Z. Luo, and H. Dai, 'Research on novel flexible high-saturation nanocrystalline cores for wireless charging systems of electric vehicles', *IEEE Trans. Ind. Electron.*, vol. 68, no. 9, pp. 8310–8320, Sep. 2021.

[25] S. Wu, C. Cai, L. Jiang, J. Li, and S. Yang, 'Unmanned aerial vehicle wireless charging system with orthogonal magnetic structure and position correction aid device', *IEEE Trans. Power Electron.*, vol. 36, no. 7, pp. 7564–7575, Jul. 2021.

[26] C. Cai, S. Wu, L. Jiang, Z. Zhang, and S. Yang, 'A 500-W wireless charging system with lightweight pick-up for unmanned aerial vehicles', *IEEE Trans. Power Electron.*, vol. 35, no. 8, pp. 7721–7724, Aug. 2020.

[27] Z. Bie, J. Zhang, K. Song, D. Wang, and C. Zhu, 'A free-rotation asymmetric magnetic coupling structure of UAV wireless charging platform with conformal pickup', *IEEE Trans. Ind. Electron.*, vol. 69, no. 10, pp. 10154–10161, Oct. 2022.

[28] S. Wu, C. Cai, X. Liu, W. Chai, and S. Yang, 'Compact and free-positioning omnidirectional wireless power transfer system for unmanned aerial vehicle charging applications', *IEEE Trans. Power Electron.*, vol. 37, no. 8, pp. 8790–8794, Aug. 2022.

4   Control for Wireless Charging of UAVs

4.1   Introduction

In response to the problem of limited UAV flight time (typically no more than 30 minutes), wireless power transfer (WPT) offers an engaging alternative to manually replacing batteries [1–4]. In-flight wireless charging of UAVs is an attractive approach to enable unmanned, flexible aerial energy delivery [5, 6] without shutting down the UAV's flight control system (unlike wireless charging after landing on a charging platform). However, while a UAV is charged while hovering, it is unavoidably influenced by environmental perturbations such as wind. This means there will be continuous fluctuation in the interactions between the coupling coils, which can dramatically degrade the energy transfer and result in destabilized power transfer or even breakdown. The weight of the UAV is also an important aspect: carrying additional control circuits or communication circuits will increase the energy loss of the UAV while hovering. Therefore, in practical applications the main issues for UAV hovering wireless charging system are (1) maintaining a stable charging output in the presence of continuous fluctuations in mutual inductance and (2) the UAV's load limitations.

Different control methods can overcome the continuous interference with the coupling strength [7–11], and thus maintain the stable transmission feature of hovering wireless charging. By adopting the coil optimization design from Chapter 3, the system offset resistance and the coupling strength can be improved to a limited extent, but the wide range of coupling variations in the hovering state of the UAV will still affect the system's output stability [12].

The general control schemes for UAV wireless charging include dual-side control [8, 13] and single-side control [5, 9–11]. An online trained radial basis function neural network (RBFNN) control algorithm has been proposed to ensure a constant output of the hovering wireless charging system for UAVs [8]. This system exhibits a faster response time and smaller steady-state error compared with common PID controls. However, it is a dual-side control requiring a wireless communication module, thus causing problems such as transmission delay or data loss.

In single-side control, the control circuits of the primary-side control [5, 9, 10] are mounted only on the primary side of the WPT system, and can meet the lightweight UAV-side requirement for in-flight charging. Several control schemes for the primary side of the UAV are described below. Figure 4.1 shows a current control scheme [5]

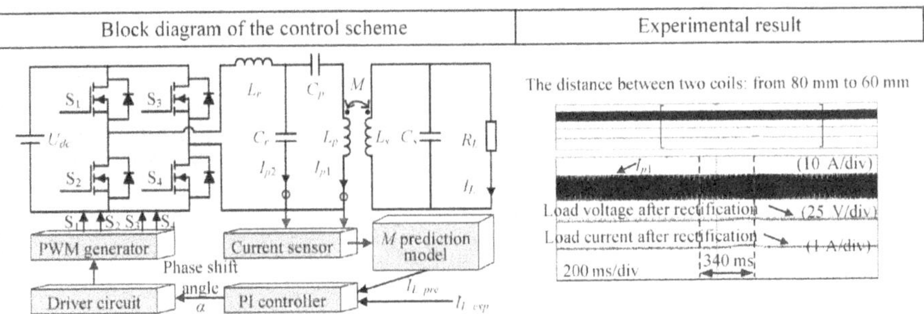

Figure 4.1 A method based on mutual inductance estimation.

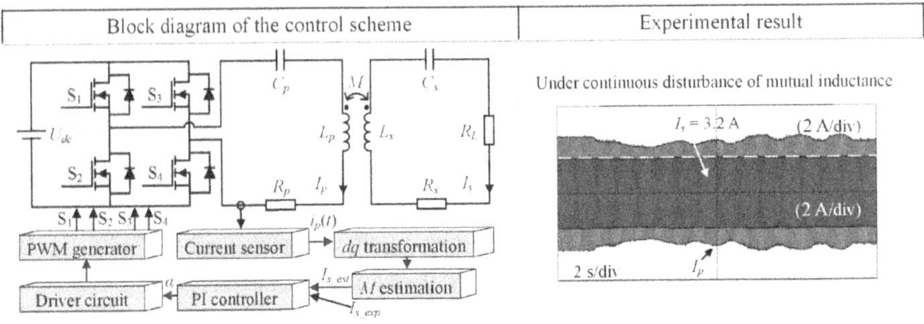

Figure 4.2 A method based on a *dq* dynamic coordinate system.

based on a high-order topology, which relies on the characteristics of the circuit topology to realize the mutual inductance estimation according to the primary side, and realizes the control of the output current based on the predicted mutual inductance. This scheme can realize a stable load current against load and mutual inductance variations, and its control circuit is only mounted on the primary side.

As shown in Fig. 4.2, an output current control scheme based on a *dq* synchronized reference coordinate system is proposed in [9], in which the use of a synchronized reference coordinate system enables decoupling of the circuit variables and thus real-time estimation of the mutual inductance is achieved through primary-side detection. The estimated mutual inductance is then used to design a constant current control algorithm to ensure a constant output as the UAV position changes. These two control methods for wireless charging of in-flight UAVs are described in detail in the rest of this chapter.

A concept based on quantum physics, known as the parity–time (PT) symmetry, was applied to the WPT systems and has gained the attention of a wide range of researchers because of the robust output power and efficiency that can be achieved despite variations in mutual inductance [14, 15]. Essentially, the PT-WPT system is realized with self-excited oscillations on the primary side to achieve negative resistance and hence stable output characteristics. Figure 4.3 depicts a novel

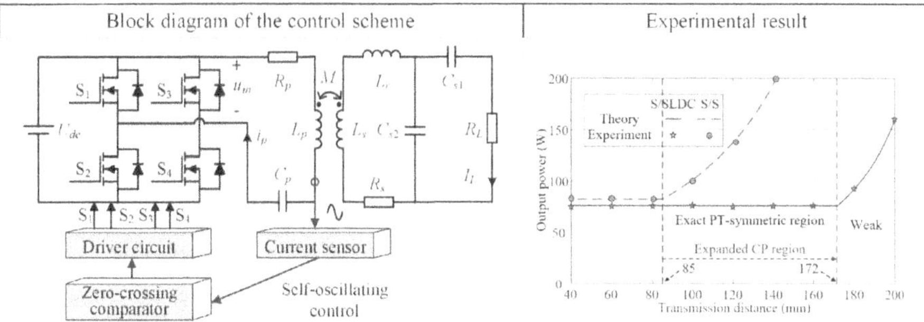

Figure 4.3 A method based on parity–time symmetry.

higher-order topology where the constant charging power range can be expanded [10]. The system implementation methodology and the relevant constant power region expansion scheme are presented in Chapters 5 and 6, respectively.

4.2　Model-Predicted Control with High-Order LCC-P Topology

Constant current (CC) charging is a common technique used to maintain reliable performance of UAV Li-ion batteries [16]. For CC charging in WPT systems, the emphasis of previous scholarly studies has included control scheme design and the proposal of topological networks to overcome the effects of coupling strength and load variations. To achieve the WPT system despite mutual inductance variations, commonly used methods include double-side control [17–20] and single-side control [21, 22]. For load perturbation, a common method is the implementation of high-order topologies [23], such as used in an inductor–capacitor–capacitor (LCC) topology network, which can automatically achieve a constant output under load variation [24].

Unlike other WPT applications [25, 26], in-flight wireless charging systems for UAVs face greater challenges. The coupling strength fluctuations of coupling coils is affected by environmental factors such as wind, which is a continuous and unknown variable. Therefore, UAV charging while hovering needs to achieve a robust output current despite continuous perturbation of the mutual inductance. In addition, considering the varied power consumption of different UAV models, the UAV must regulate its power use to allow adequate power replenishment. A lightweight UAV design is especially important because any extra weight on the UAV increases the power consumption by hovering. To summarize: to achieve stable in-flight wireless charging for UAVs, three key issues must be addressed: continuous variations of the coupling strength, expected output variations, and weight constraints.

In this section, a mutual inductance prediction scheme with the LCC-P high-order topology is presented to realize CC output control for UAV hovering charging. The high-order circuit topology can realize CC output control despite battery load variation, while the secondary-side parallel compensation satisfies the demand for lightweight design on the pickup side. This method can realize both the prediction of mutual inductance with the detection of transmitting current and the flexible regulation of output current according to the predicted mutual inductance, which improves system robustness. Compared with other control strategies [17], this scheme does not require new control devices on the receiving side, meaning the UAV weight (and so power consumption) is not increased.

4.2.1   System Modelling and Analysis

Figure 4.4 shows a high-order LCC-P topology for a hovering wireless charging system, where V_{dc} denotes the given voltage and S_1–S_4 form a full-bridge inverter. L_p is the transmitter coil, L_s is the pickup coil, and M is the mutual inductance of the two coils in the WPT system. L_r stands for the resonance inductance, and C_p, C_s, and C_r represent the compensated resonance of the corresponding coils, respectively. R_r, R_p, and R_s represent the internal resistance of the corresponding coils, and R_L represents the system equivalent impedance of the UAV battery R_{bat} in the WPT system. Here, $R_L = 8R_{bat}/\pi^2$.

Figure 4.4(b) illustrates the T-model circuit of the LCC-P topology. The corresponding circuit model is obtained using Kirchhoff's voltage law:

$$I_{in} = I_{p1} + I_{p2}, \tag{4.1}$$

$$U_{in} = (j\omega L_r + R_r)I_{in} + \frac{1}{j\omega C_r}I_{p2}, \tag{4.2}$$

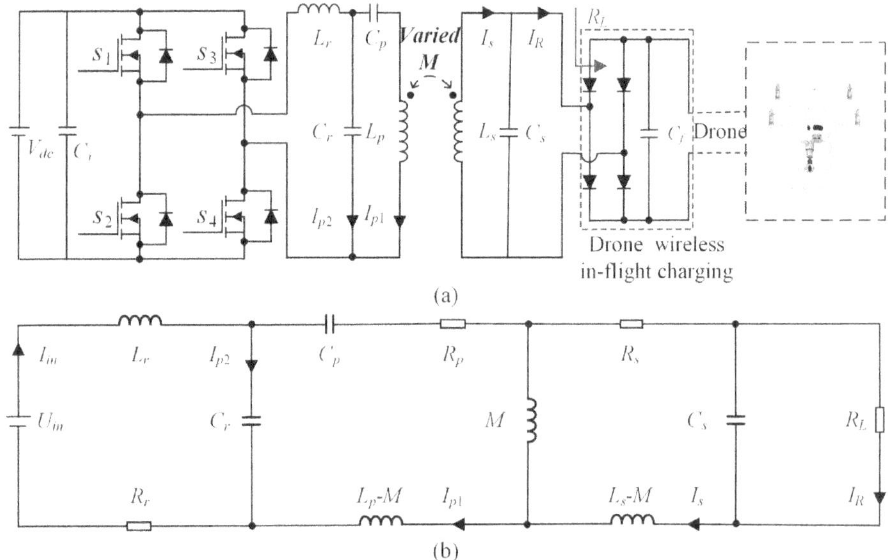

Figure 4.4 (a) System architecture with high-order topology; (b) circuit for the equivalent T-model.

$$\frac{1}{j\omega C_r}I_{p2} = \left(\frac{1}{j\omega C_p} + j\omega L_p + R_p\right)I_{p1} - j\omega M I_s, \tag{4.3}$$

$$j\omega M\left(I_{p1} - I_s\right) = \left(j\omega(L_s - M) + R_s + \frac{R_L}{j\omega C_s R_L + 1}\right)I_s. \tag{4.4}$$

The impedances Z_1 and Z_2 satisfy that $Z_1 = 1/j\omega C_p + j\omega L_p + R_p$ and $Z_2 = j\omega L_s + R_s + R_L/(j\omega C_s R_L + 1)$. The system operating frequency is represented by ω, and the system given voltage is denoted by U_{in}. In order to ensure the resonance of the capacitor with the inductor at the desired frequency, the compensation capacitor is designed as the following value:

$$C_r = \frac{1}{\omega^2 L_r}, \quad C_p = \frac{1}{\omega^2 L_p}, \quad C_s = \frac{1}{\omega^2 L_s}. \tag{4.5}$$

Based on Eqs. (4.1)–(4.5), the receiving coil current I_s and load current I_R are expressed as:

$$I_s = \frac{j\omega M U_{in}}{j\omega C_r R_r\left(\omega^2 M^2 + Z_1 Z_2\right) + Z_2(j\omega L_r + R_r)}, \tag{4.6}$$

$$I_R = \frac{j\omega M U_{in}/(j\omega C_s R_L + 1)}{j\omega C_r R_r\left(\omega^2 M^2 + Z_1 Z_2\right) + Z_2(j\omega L_r + R_r)}. \tag{4.7}$$

The coil internal resistors R_r, R_p, and R_s in a WPT system are normally small and negligible. Therefore, I_R is calculated as:

$$I_R = \frac{M U_{in}}{j\omega L_r L_s}. \tag{4.8}$$

Based on Eq. (4.8), current I_R is not affected by R_L but by M. The system load power and system efficiency are obtained as:

$$P_{out} = I_R^2 R_L = \frac{M^2 U_{in}^2 R_L}{\omega^2 L_r^2 L_s^2},$$

$$\eta_o = \frac{P_{out}}{U_{in} I_{in}} = \frac{M^2 R_L}{\omega L_s^2} \Big/ \sqrt{\left(L_r L_s + M^2\right)^2 + \frac{M^4 R_L^2}{\omega^2 L_s^4}}. \tag{4.9}$$

Currents I_{p1} and I_s are given by:

$$I_{p1} = \frac{U_{in}}{j\omega L_r}, \tag{4.10}$$

$$I_s = \frac{M U_{in}}{j\omega L_r L_s}\left(1 + j\frac{R_L}{\omega L_s}\right). \tag{4.11}$$

Another primary current, I_{p2}, can be obtained as:

$$I_{p2} = \omega^2 C_r M I_s = \frac{M^2 U_{in}}{j\omega L_r^2 L_s}\left(1 + j\frac{R_L}{\omega L_s}\right). \tag{4.12}$$

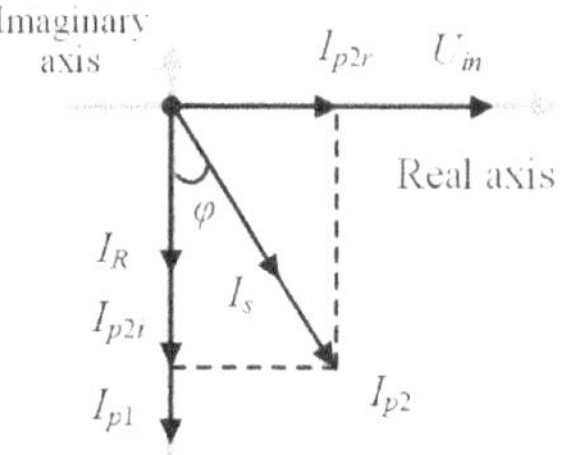

Figure 4.5 The system state variable phasor diagram.

According to Eqs. (4.10) and (4.12), the formulation of mutual inductance is expressed as:

$$M = \sqrt{\frac{\left|I_{p2}L_rL_s\right|}{\left|I_{p1}\right| \cdot \left|1 + j\frac{R_L}{\omega L_s}\right|}}. \tag{4.13}$$

Therefore, when the system parameters such as inductors and load are given in advance, mutual inductance could be estimated in real time by measuring the currents I_{p1} and I_{p2}.

Mutual inductance prediction is challenging and important during UAV charging, when the battery load is changing and cannot be measured. Figure 4.5 depicts the phase diagram of the state variable on the transmitter and receiver in the WPT system to explain the relationship between them. Here, $I_{p2i} = I_{p2}\cos\varphi = M^2 U_{in} /j\omega L_r^2 L_s$. After this analysis of the relationship between current and voltage in the circuit, the mutual inductance could also be expressed with the transmitting current:

$$M = \sqrt{\frac{\left|I_{p2}\cos\varphi\right|}{\left|I_{p1}\right|}L_rL_s}. \tag{4.14}$$

Therefore, for a WPT system with LCC-P high-order network, if the phase difference φ and magnitudes I_{p1} and I_{p2} are obtained, mutual inductance could be estimated when the value of the load is unknown and is changing. Based on Eq. (4.14), real-time mutual inductance estimation under unknown loads could be realized, which is especially suitable for scenarios where the loads are constantly changing during the charging process and are not easy to measure.

According to Eq. (4.8), the change in mutual inductance will cause the load current to change, while the load change will not affect it. Especially in the hover charging scenario, the inter-coil coupling will continuously change, which in turn affects the system output. Hovering UAVs have lightweight design requirements on the receiver side of the WPT system, as well as changing desired charging current requirements due to power loss and the environment.

4.2.2　Mutual Inductance Prediction and Controller Design

To address the challenges specific to UAV in-flight wireless charging mentioned above, this section presents a CC control scheme combined with mutual inductance estimation. First, the mutual inductance prediction could be realized according to Eq. (4.13) in the WPT system under the LCC-P high-order topology. Using the known parameters, mutual inductance could be estimated by the measurement of the amplitudes of transmitting currents I_{p1} and I_{p2}. Then, according to Eq. (4.8), load current I_R can also be predicted accurately and easily. If the load information cannot be measured, the mutual inductance could still be estimated by only detecting φ between the measured currents I_{p1} and I_{p2} based on Eq. (4.14). Also, considering the possibility that the circuit may not be fully resonant in a realistic scenario, we can theoretically compensate for the effect of detuning on the mutual inductance prediction. Unlike some other schemes, the mutual inductance estimation scheme presented here is better in real time due to the fact that it is implemented in the transmitter. Utilizing the real-time estimated mutual inductance, an output CC control scheme is further proposed here to avoid current fluctuations caused by system parameter variations and to ensure system robustness.

Figure 4.6 describes the CC control algorithm based on the mutual inductance dynamic prediction proposed here. The primary-side LCC topology compensation and the secondary-side parallel compensation are used here. The prediction of mutual inductance based on Eq. (4.13) can be achieved by measuring the RMS values of I_{p1} and I_{p2} using sensors at the primary side. The output current value I_{R_pre} can be predicted based on Eq. (4.8). When the desired output current is designed in the controller, the phase-shift angle α of the inverter will be adjusted, which in turn adjusts the inverter output voltage, realizing the tracking and regulation of system output current.

As shown in Fig. 4.7, the dynamically predictive CC control algorithm consists of two main parts, known as real-time mutual inductance estimation and

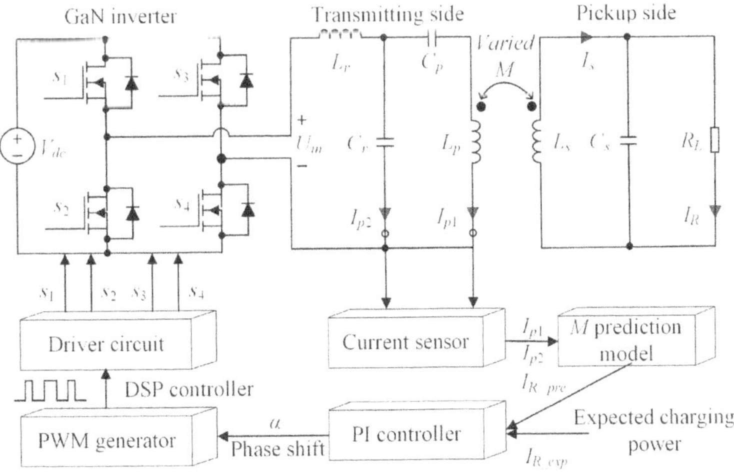

Figure 4.6 A system control block diagram.

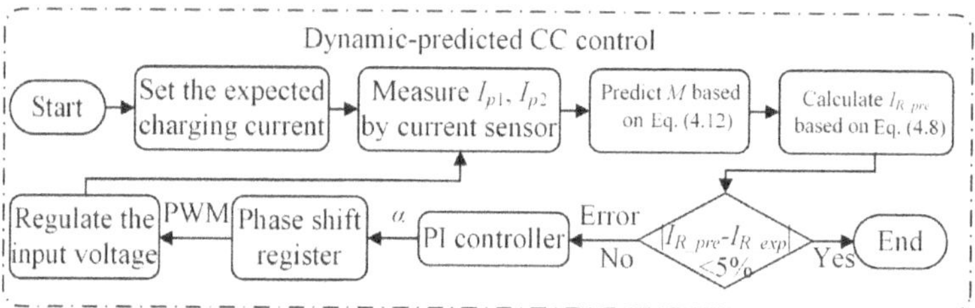

Figure 4.7 An algorithm flowchart.

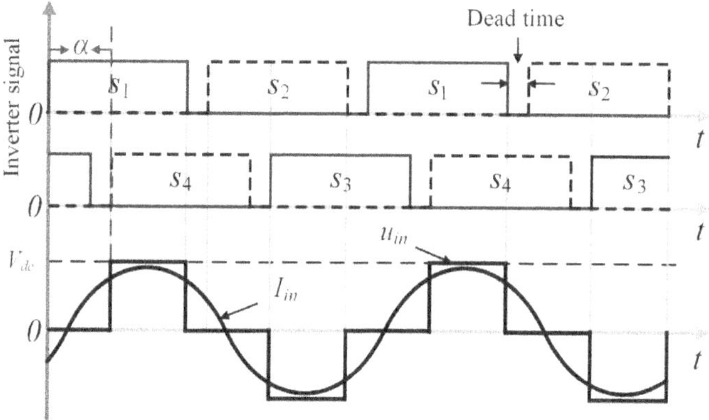

Figure 4.8 A schematic diagram of drive signals and inverter outputs.

phase-shift-based current control. During the prediction of mutual inductance, the current is measured and controlled only at the transmitter side, avoiding the addition of detection and control circuits in the receiver and mitigating the extra UAV-side power loss due to the secondary-side device. Considering the practical requirements of a UAV in-flight charging system, a phase-shift-based output current control scheme is adopted here. The above high-performance controller-based control scheme can achieve rapid output tracking against mutual inductance variations and also ensure the UAV-side lightweight framework.

Figure 4.8 depicts the inverter switching signals with phase-shift control. It can be seen that as angle α increases, the inverter output voltage decreases and ensures that the output current I_R is tracked and adjusted. Here, the dead time can be ignored because it is short compared to the switching time of the switching tubes [27]. Then, the system input voltage U_{in} is represented as:

$$U_{in} = \frac{2\sqrt{2}}{\pi} V_{dc} \cos\frac{\alpha}{2}. \tag{4.15}$$

Therefore, load current I_R with the control strategy is represented as:

$$|I_R| = 2\sqrt{2}V_{dc}M\cos\frac{\alpha}{2}/\pi\omega L_r L_s. \tag{4.16}$$

4.2.3   Simulation Verification

To validate the effectiveness of the mutual inductance estimation and the CC control, a simulation model was built [5]. We designed the simulated validation of mutual inductance prediction by ignoring and including the coil internal resistance.

As shown in Fig. 4.9, the mutual inductance range of coupling coefficients from 0.06 to 0.25 was experimentally tested to prove the accuracy of the proposed estimation method. Figures 4.9(a) and (b) depict the prediction results of mutual inductance under different loads when the coil internal resistor is ignored. It can be seen that the estimation is more accurate, with an error rate of less than 5.5%. Significantly, at high coupling coefficients (i.e. between 0.08 and 0.25) the error rate is less than 2%. Figures 4.9(c) and (d) depict the estimation effect for different values of load when the coil internal resistance is included. Compared to the previous simulation results, the prediction error rate is higher but still below 5%. In summary, the estimation accuracy of mutual inductance is more than 95% for UAV hover charging

In the simulation model, the proposed CC control algorithm was validated using software to simulate a step variation in coupling strength. The output current waveforms with varied mutual inductance and different system load values can be seen in Fig. 4.10. Coupling coefficients and the error rate of mutual inductance prediction are shown at the top.

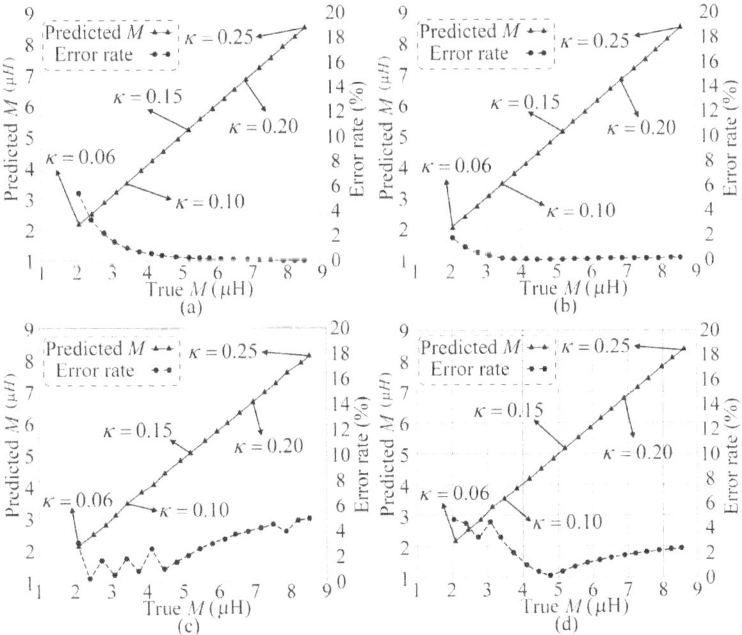

Figure 4.9 Estimated simulation results. (a) Neglecting internal resistors with $R_L = 10\ \Omega$. (b) Neglecting internal resistors with $R_L = 20\ \Omega$. (c) Including internal resistors with $R_L = 10\ \Omega$. (d) Including internal resistors with $R_L = 20\ \Omega$.

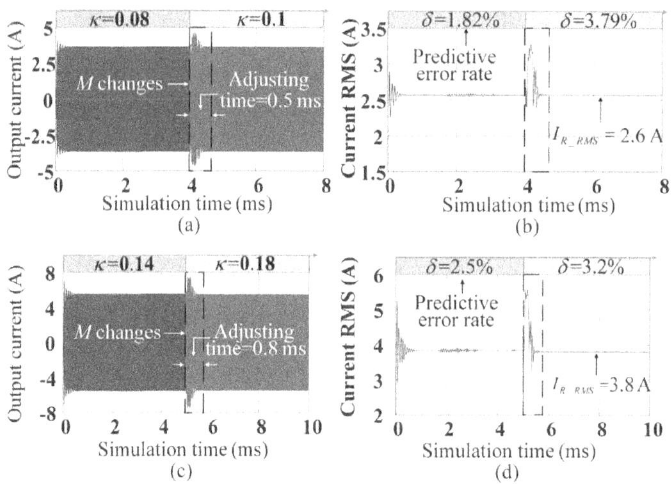

Figure 4.10 Simulated control results. (a) Load current with $R_L = 10\ \Omega$; (b) RMS value of load current with $R_L = 10\ \Omega$; (c) load current with $R_L = 20\ \Omega$; (d) RMS value of load current with $R_L = 20\ \Omega$.

Figures 4.10(a) and (b) show the load current waveforms of the system for a load and desired output current of 10 Ω and 2.6 A, respectively. It can be seen that the current regulation time is around 0.5 ms, with an error rate of around 3.79%. The current output waveforms for the load and expected current of 20 Ω and 3.8 A, respectively, illustrated in Fig. 4.10(c) and (d), show that its tuning time is around 0.8 ms, with an error rate of 3.2%. These simulations indicate that the CC control method with mutual inductance prediction can realize real-time current tracking under mutual inductance changes, which ensures the UAV hover charging can cope with the environment-induced changes in relative position of coupling coils.

4.2.4 Experiment Verification

We built the WPT system with the LCC-P compensation network as illustrated in Fig. 4.11 to verify this proposed mutual inductance prediction and the CC control algorithm. The model Itech IT6722A power supply with GaN full-bridge inverter generates the 100 khz AC power required for the WPT system.

From this experimental validation, the predicted and actual values of mutual inductance under various conditions are shown in Fig. 4.12, where it is clear that the estimated values are very similar to the actual values. Considering the possible resonance shift of the system (usually within 10%), Fig. 4.13 shows that the mutual inductance can be estimated with a precision of greater than 95%. In summary, this proposed prediction method is highly accurate, which is good news for the design of the CC control algorithm for the UAV in-flight charging system.

In these experimental results, the prediction error is larger when the transmission distance is larger, which is mainly dependent on the detection capability of the current

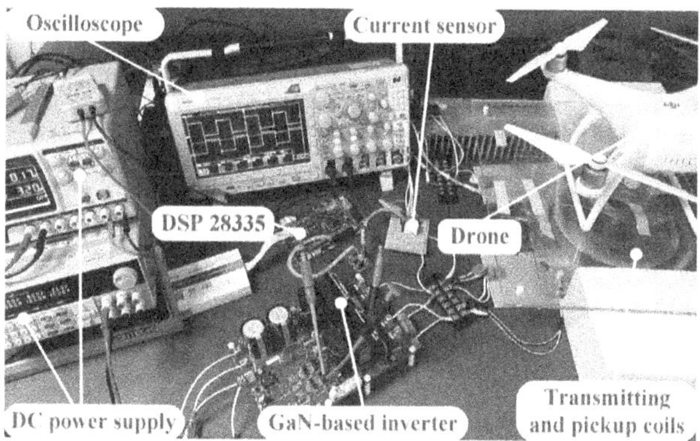

Figure 4.11 Experimental prototype.

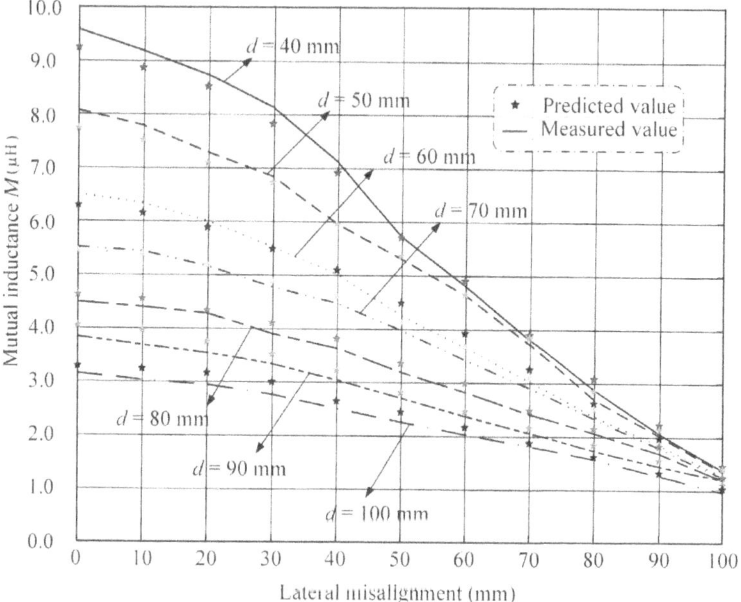

Figure 4.12 Real and predicted values of mutual inductance with misalignments and transfer distances.

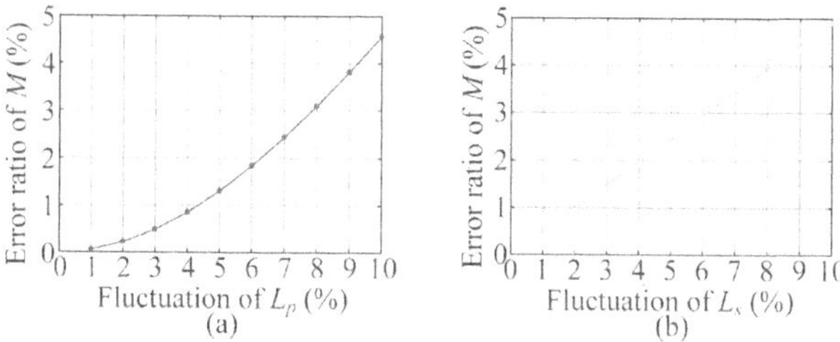

Figure 4.13 Error rate against the variation of (a) L_p, (b) L_s.

sensor in the experiment. This is because system transmission performance is weaker when coupling strength is weaker; therefore, the transmitting-side current measured by the sensor is smaller. Small errors in sensor detection in the low-current case can drastically affect the mutual inductance prediction accuracy. It is worth noting that the above long-distance transmission scenario with high error has a low probability of occurring in the UAV in-flight charging system. The UAV should be closer to the transmitter coil for high performance charging in real scenarios. The above proposed scheme of mutual inductance prediction does not require a known load value, it only needs to be detected and controlled at the primary side, and can realize more than 92.5% mutual inductance prediction accuracy within the coupling coefficient of the UAV in-flight charging system, which is a big advantage.

The experimental cases are designed to test the practicability of the proposed CC control algorithm for mutual inductance prediction. The experimental conditions are:

- DC voltage $= 10$ V
- load $= 25\ \Omega$
- natural resonant frequency $f = 100$ kHz.

The coupling coils are initially aligned, with an air gap of 75 mm, which is subsequently changed to 55 mm with increased coupling strength. With the change in mutual inductance, the system voltage, current, and drive signal of the inverter are as shown in Fig. 4.14. It can be seen that by adjusting the phase angle to 63° using the proposed current regulation algorithm, the system load current will be kept at the desired 0.8 A. At this point, the output power and system efficiency are 16 W and 84.21%, respectively. Figure 4.15 shows the experimentally obtained primary current I_{p1} and the system load voltage and current. The results show that the mutual inductance is predicted with high accuracy under a changing relative coil position, and the current can be adjusted to the desired value within 300 ms. The control regulation time of the actual system is shorter because the mutual inductance variation is slower due to the manual movement of the coil used in the experiment.

In this section, the CC regulation algorithm integrated with the real-time estimation of mutual inductance under the LCC-P high-order topology is presented to realize output current regulation with a lightweight receiver side. To address the challenges

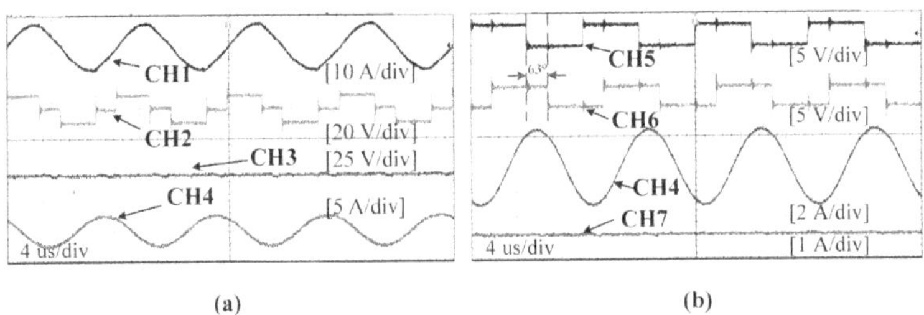

(a)   (b)

Figure 4.14 Experimental waveforms (a) CH1 is I_{in}; CH2 is U_{in}; CH3 is U_R; CH4 is I_{p1}; (b) CH5 is S_1; CH6 is S_4; CH4 is I_{p1}; CH7 is I_R.

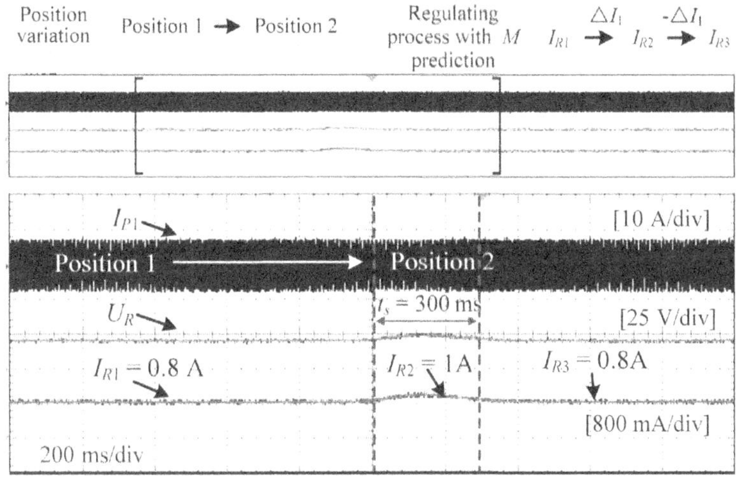

Figure 4.15 Experimental waveforms of I_{p1}, U_R, and I_R when mutual inductance varies due to positional change.

specific to UAV hovering wireless charging, here the real-time estimation of mutual inductance is realized with detection on the primary-side current only. The adjustment of the desired load current can be achieved in integration with the phase-shift-control algorithm. The above control scheme does not require a communication link, which enhances the real-time performance, and its stability and robustness are verified through experiments. Therefore, the use of the CC control algorithm in UAV hovering charging will maintain a stable charging performance of the UAV and extend the its operational time.

4.3 Rotating-Coordinate-Based Mutual Inductance Estimation

Hovering wireless charging is a highly promising application-oriented UAV charging technology, and is regarded as an ideal solution to cope with the major constraint of short range and running duration of electrically powered UAVs. However, UAV positioning deviations and irregular environmental perturbations inevitably lead to constant changes in the relative positions of the coupling coils, with resulting mutual inductive continuity perturbations that substantially reduce the power transfer capacity. In this section, a coupling strength estimation technique using dq-based dynamic modelling is introduced [9], which can predict changes in coupling strength in a fast, real-time, and accurate manner, and thus provide the prerequisites for the realization of stable energy transmission. By using a series–series compensation topology and deploying it on the transmitter side, this method meets the lightweight design requirements of the receiver side. By applying a dq dynamic model, the method is able to achieve fast tracking of continuity perturbations in the coupling strength, without a communication link. The test results show that this mutual inductance estimation scheme could effectively deal with coupling strength continuity perturbations in the hovering wireless charging system, and so provide a feasible solution to facilitate in-flight wireless charging technology.

4.3.1   System Modelling of *dq* Synchronous Reference Frame Model

The circuit schematic of the series–series type WPT system is shown in Fig. 4.16, where the transmitter circuit consists of the coil L_t, internal resistance R_t, and compensation capacitor C_t, and the receiver circuit consists of the coil L_r, internal resistance R_r, compensation capacitor C_r, and the load R_L. M denotes the coupled mutual inductance. The system input voltage u_s, current i_t, and load current i_r can be expressed as:

$$u_s = U_s \cos(\omega_0 t), \tag{4.17}$$

$$i_t = I_t \cos(\omega_0 t - \varphi_t), \tag{4.18}$$

$$i_r = I_r \cos(\omega_0 t - \varphi_r), \tag{4.19}$$

where U_s, I_t, and I_r are the amplitudes of the input voltage, transmitter, and receiver currents, respectively. ω_0 represents system operating frequency. φ_t and φ_r are the hysteresis phases of the primary and pickup currents with respect to the input voltage, respectively. Figure 4.16 depicts the equivalent circuit of the actual WPT system, which is regarded as an α-axis model in the static coordinate system. This actual system is lagged by 90 degrees as a β-axis model, as depicted in Fig. 4.17. We multiply the unit imaginary numbers with the voltages and currents in Fig. 4.17 and use Euler's formula – that is, synthesize the α-axis model with the β-axis model, where the current is used as an example, to derive the following equations:

$$I_t \cos(\omega_0 t - \varphi_t) + jI_t \sin(\omega_0 t - \varphi_t) = I_{\alpha\beta t} = I_t e^{j(\omega_0 t - \varphi_t)} = I_{dqt} e^{j\omega_0 t}, \tag{4.20}$$

$$I_r \cos(\omega_0 t - \varphi_r) + jI_r \sin(\omega_0 t - \varphi_r) = I_{\alpha\beta r} = I_r e^{j(\omega_0 t - \varphi_r)} = I_{dqr} e^{j\omega_0 t}, \tag{4.21}$$

where I_{dqt} and I_{dqr} stand for $I_t e^{-j\varphi_t}$ and $I_r e^{-j\varphi_r}$, respectively.

$I_{\alpha\beta t}$ and $I_{\alpha\beta r}$ obtained after synthesis are the rotation vectors in the $\alpha\beta$ stationary coordinate system. The circuit model of the $\alpha\beta$ stationary coordinate system obtained after synthesis is given in Fig. 4.18. Voltage $U_{\alpha\beta Lt}$ across the transmitter-side inductor in the $\alpha\beta$ stationary coordinate system can be deduced as:

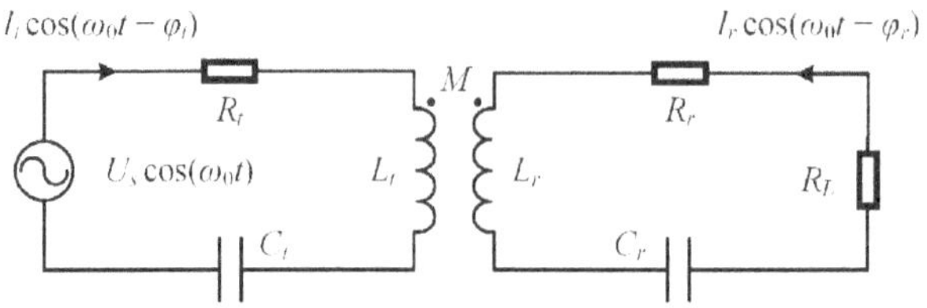

Figure 4.16 System equivalent circuit.

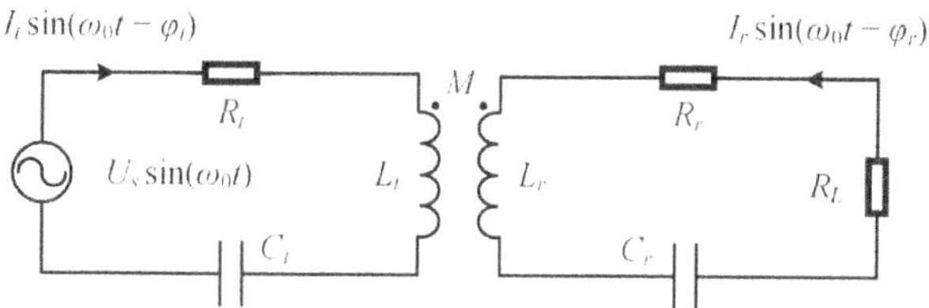

Figure 4.17 Imaginary axis equivalent circuit.

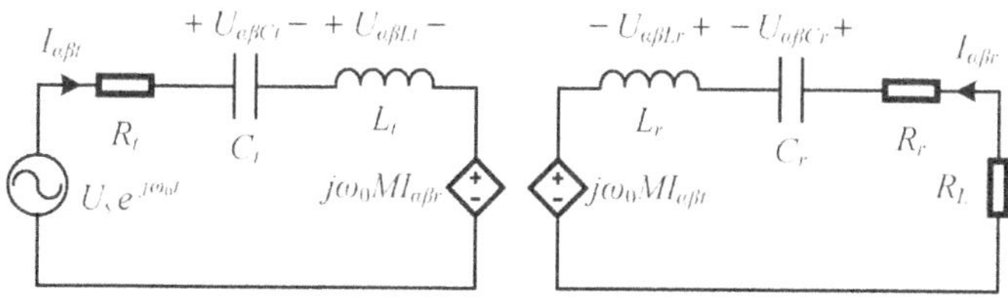

Figure 4.18 Static coordinate system circuit model.

$$U_{\alpha\beta Lt} = L_t \frac{dI_{\alpha\beta t}}{dt} = L_t \frac{dI_{dqt}e^{j\omega_0 t}}{dt}$$

$$= \left[L_t \frac{dI_{dqt}}{dt} + j\omega_0 L_t I_{dqt} \right] e^{j\omega_0 t} = U_{dqLt}e^{j\omega_0 t}, \tag{4.22}$$

where $e^{j\omega_0 t}$ represents the rotation factor, which describes the sinusoidal variation of the inductor voltage with angular frequency ω_0 in the time domain over time, and $|U_{dqLt}|$ represents the amplitude variation of the sinusoidal waveform. In the $\alpha\beta$ static coordinate system, $U_{\alpha\beta Lt}$ could be considered as a rotation vector with mode $|U_{dqLt}|$ and angular velocity ω_0. The voltage $U_{\alpha\beta Lr}$ across the receiver-side inductor could be deduced in the same way:

$$U_{\alpha\beta Lr} = \left[L_r \frac{dI_{dqr}}{dt} + j\omega_0 L_r I_{dqr} \right] e^{j\omega_0 t}. \tag{4.23}$$

In addition, voltage $U_{\alpha\beta Ct}$ across the primary-side capacitor can be expressed as:

$$U_{\alpha\beta Ct} = \frac{1}{C_t} \int I_{dqt} e^{j\omega_0 t} dt$$

$$= \frac{1}{j\omega_0 C_t} \left[I_{dqt} e^{j\omega_0 t} - \frac{1}{j\omega_0} \frac{dI_{dqt}}{dt} e^{j\omega_0 t} + \frac{1}{j\omega_0} \int \frac{d^2 I_{dqt}}{dt^2} e^{j\omega_0 t} dt \right]. \tag{4.24}$$

After making an engineering approximation and neglecting higher-order nonlinear terms, the derivation is:

$$U_{\alpha\beta Ct} = \left[\frac{1}{\omega_0^2 C_t} \frac{dI_{dqt}}{dt} + \frac{1}{j\omega_0 C_t} I_{dqt} \right] e^{j\omega_0 t}. \tag{4.25}$$

The voltage $U_{\alpha\beta Cr}$ across the receiver-side capacitor can be derived as:

$$U_{\alpha\beta Cr} = \left[\frac{1}{\omega_0^2 C_r}\frac{dI_{dqr}}{dt} + \frac{1}{j\omega_0 C_r}I_{dqr}\right]e^{j\omega_0 t}. \tag{4.26}$$

Then, the $\alpha\beta$ stationary coordinate system model could be derived as:

$$U_s e^{j\omega_0 t} = \left[R_t I_{dqt} + \left(L_t + \frac{1}{\omega_0^2 C_t}\right)\frac{dI_{dqt}}{dt} + j\left(\omega_0 L_t - \frac{1}{\omega_0 C_t}\right)I_{dqt} + j\omega_0 M I_{dqr}\right]e^{j\omega_0 t}, \tag{4.27}$$

$$0 = \left[(R_r + R_L)I_{dqr} + \left(L_r + \frac{1}{\omega_0^2 C_r}\right)\frac{dI_{dqr}}{dt} + j\left(\omega_0 L_r - \frac{1}{\omega_0 C_r}\right)I_{dqr} + j\omega_0 M I_{dqt}\right]e^{j\omega_0 t}. \tag{4.28}$$

The state variables in this system could all be treated as rotational vectors with the same angular velocity ω_0 in the static coordinate system.

Figure 4.19(a) gives a schematic diagram of the $\alpha\beta$ stationary coordinate system and the rotation vector as an example of $I_{\alpha\beta t}$. Given that the current and voltage in the $\alpha\beta$ stationary coordinate system model are both rotation vectors with angular velocity ω_0, a dq rotation coordinate system with angular velocity ω_0, as depicted in Fig. 4.19(b), is established on the basis of the $\alpha\beta$ stationary coordinate system using the Park transformation for coordinate transformation. As a consequence, the rotation vectors in the static coordinate system may be expressed as static vectors in the dq rotated coordinate system. The conversion process can be performed through multiplication of Eqs. (4.27) and (4.28) with $e^{-j\omega_0 t}$. Hence, the dq dynamic model is given as:

$$U_{dqs} = R_t I_{dqt} + \left(L_t + \frac{1}{\omega_0^2 C_t}\right)\frac{dI_{dqt}}{dt} + j\left(\omega_0 L_t - \frac{1}{\omega_0 C_t}\right)I_{dqt} + j\omega_0 M I_{dqr}, \tag{4.29}$$

$$0 = (R_r + R_L)I_{dqr} + \left(L_r + \frac{1}{\omega_0^2 C_r}\right)\frac{dI_{dqr}}{dt} + j\left(\omega_0 L_r - \frac{1}{\omega_0 C_r}\right)I_{dqr} + j\omega_0 M I_{dqt}, \tag{4.30}$$

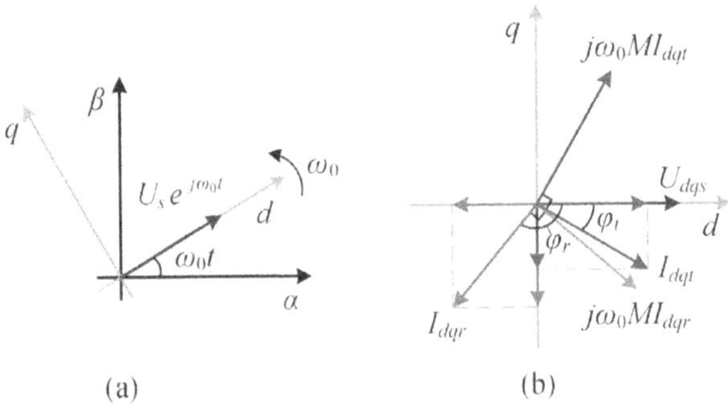

Figure 4.19 (a) The relationship between the dq coordinate system and the $\alpha\beta$ coordinate system. (b) A vector diagram of state variables.

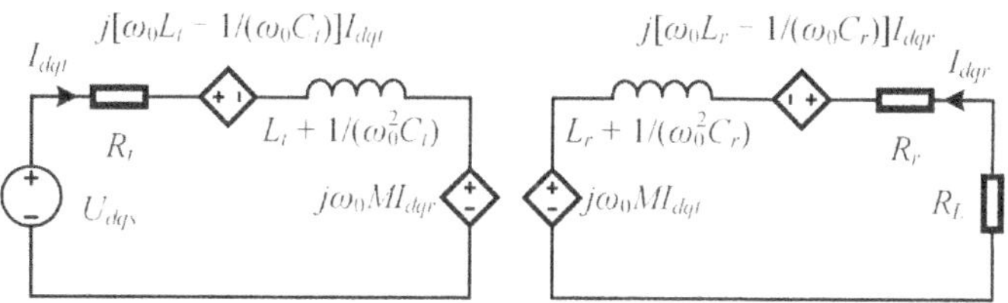

Figure 4.20 The dq circuit model.

where U_{dqs} is a non-rotating vector in the dq dynamic system, where its mode is U_s and its phase angle is 0. This dynamic model of the WPT system can be helpful when implementing control strategies with high real-time capability. Figure 4.20 illustrates the series–series compensated dq dynamic circuit model of the WPT system based on Eqs. (4.29) and (4.30).

I_{dqt} and I_{dqr} are decomposed orthogonally in the dq rotated coordinate system as follows:

$$I_{dqt} = I_{dt} + jI_{qt}, \tag{4.31}$$

$$I_{dqr} = I_{dr} + jI_{qr}, \tag{4.32}$$

where I_{dt} and I_{dr} are the values of the d-axis currents on the transmitting and pickup sides, respectively, and I_{qt} and I_{qr} represent the q-axis currents on the transmitter and receiver sides, respectively. The real part of the equation in this model belongs to the d-axis and the imaginary part to the q-axis. The transmitter side of the system is decomposed orthogonally and the equations for the real and imaginary parts are expressed, respectively, as:

$$U_s = \left|U_{dqs}\right| = R_t I_{dt} + \left(L_t + \frac{1}{\omega_0^2 C_t}\right)\frac{dI_{dt}}{dt} - \left(\omega_0 L_t - \frac{1}{\omega_0 C_t}\right)I_{qt} - \omega_0 M I_{qr}, \tag{4.33}$$

$$0 = R_t I_{qt} + \left(L_t + \frac{1}{\omega_0^2 C_t}\right)\frac{dI_{qt}}{dt} + \left(\omega_0 L_t - \frac{1}{\omega_0 C_t}\right)I_{dt} + \omega_0 M I_{dr}. \tag{4.34}$$

When the system is in non-resonance matching, the modes and phase angles of the primary and secondary currents can be denoted as $\left|I_{dqt}\right|$, $\left|I_{dqr}\right|$ and $-\varphi_t$, $-\varphi_r$, respectively. $\left|I_{dqt}\right|$ and $\left|I_{dqr}\right|$ can be obtained as:

$$I_t = \left|I_{dqt}\right| = \sqrt{I_{dt}^2 + I_{qt}^2}, \tag{4.35}$$

$$I_r = \left|I_{dqr}\right| = \sqrt{I_{dr}^2 + I_{qr}^2}. \tag{4.36}$$

The hysteresis phases of the transmit and receive currents can be denoted, respectively, as:

$$\varphi_t = -\arctan\left(\frac{I_{qt}}{I_{dt}}\right), \tag{4.37}$$

$$\varphi_r = \pi - \arctan\left(\frac{I_{qr}}{I_{dr}}\right). \tag{4.38}$$

It is easy to see that when the d-axis and q-axis current vectors in the dq dynamic coordinate system are obtained, the amplitude and phase of the transmit and receive currents can be calculated accordingly.

4.3.2 $\alpha\beta$ to dq Transformation and Mutual Inductance Estimation

The proposed dq dynamic model estimates the first-order dynamic characteristics of the system better than other common circuit models. Therefore, the model has the advantage of good real-time performance when used for parameter identification or system control. Here we provide the implementation procedure and use it to estimate the system mutual inductance. First, the circuit is considered as an α-axis current. The β-axis current is obtained by using a digital signal processor at 90 degrees. In a real system, when we measure the instantaneous value of the transmitter current, we can use the current value as the α-axis signal and then use the current value 90 degrees behind as the β-axis signal – that is, use the current before a one-quarter system cycle as the β-axis signal. Together, the two signals form a rotation vector in a static coordinate system.

Based on the derivation in the previous section, in order to convert the rotational vectors in the $\alpha\beta$ stationary coordinate system to non-rotational vectors in the dq rotational coordinate system, the multiplier $e^{-j\theta}$ is introduced, where θ is defined as:

$$\theta = \omega_0 t. \tag{4.39}$$

The introduction of this multiplier is based on the premise that system input voltage is defined in a cosine function. However, in general, for practical measurement and control purposes, the input voltage is usually defined as a sinusoidal function form:

$$u_s = U_s \sin(\omega_0 t) = U_s \cos\left(\omega_0 t - \frac{\pi}{2}\right). \tag{4.40}$$

As shown in Fig. 4.21, in this case the multiplier $e^{-j\theta}$ will position the signal that should be located on the d-axis in the dq rotational coordinate system on the q-axis. Therefore, in order to obtain the vector relationship of the axes depicted in Fig. 4.19 (b), one would need to add a positive 90 degrees to the multiplier employed.

The updated multiplier is $e^{j(-\theta+\pi/2)}$, on the basis of which the transmit current is illustrated as an example in Eq. (4.41):

$$I_{dqt} = I_{\alpha\beta t}\, e^{j\left(-\theta+\frac{\pi}{2}\right)}. \tag{4.41}$$

This equation is expressed in matrix form to specify the relationship between the vector components in the $\alpha\beta$ stationary coordinate system and the vector components in the dq rotating coordinate system – that is, the dq transformation:

$$\begin{bmatrix} I_{dt} \\ I_{qt} \end{bmatrix} = \begin{bmatrix} \sin\theta & -\cos\theta \\ \cos\theta & \sin\theta \end{bmatrix} \begin{bmatrix} I_{\alpha t} \\ I_{\beta t} \end{bmatrix}. \tag{4.42}$$

Here, $I_{\alpha t}$ and $I_{\beta t}$ denote the transient values of transmitting current.

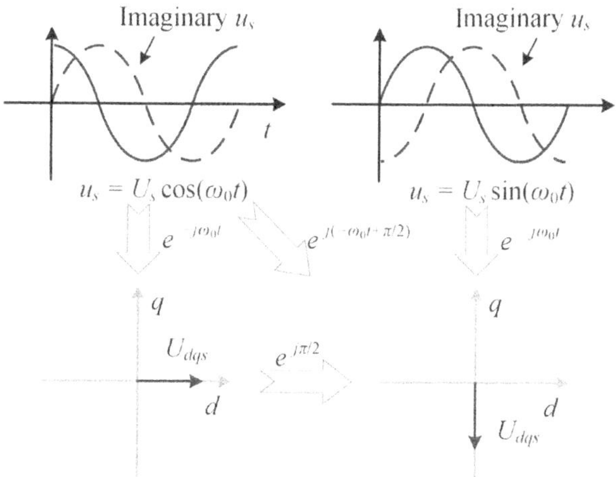

Figure 4.21 Updating the $\alpha\beta$ to dq transformation.

In order to ensure the validity of the current transient value measurement – that is, to ensure as much as possible that the transmit current is not distorted – the phase-shift control is selected to regulate the system input voltage. Based on the Fourier series expansion, the relationship between basic wave content U_s and angle α of u_{in} under phase-shift control is deduced:

$$U_s = \frac{2}{\pi}\int_{\frac{\alpha}{2}}^{\pi-\frac{\alpha}{2}} U_{dc}\sin(\theta)d\theta$$
$$= \frac{4}{\pi}U_{dc}\cos\left(\frac{\alpha}{2}\right). \tag{4.43}$$

Here, U_{dc} denotes the system input voltage. Based on Eq. (4.43), voltage U_s could be calculated directly with the known phase-shift angle, which indicates that the input voltage magnitude information can be accurately obtained in the digital signal processor without the use of a voltage sensor, which is another advantage of using phase-shift control.

According to the obtained dynamic model, the estimation scheme for mutual inductance can be achieved with the detection of the primary current. The scheme consists of the following: I_{dt} and I_{qt} are determined from the primary current measured in real time using the dq transform. The input voltage amplitude U_s can be calculated from the known phase-shift angle and utilizing Eq. (4.43). Thus, with the circuit equations of Eqs. (4.33) and (4.34), the following equation can be obtained:

$$U_{dMt} = -\omega_0 M I_{qr} = U_s - R_t I_{dt} - \left(L_t + \frac{1}{\omega_0^2 C_t}\right)\frac{\Delta I_{dt}}{\Delta t} + \left(\omega_0 L_t - \frac{1}{\omega_0 C_t}\right)I_{qt}, \tag{4.44}$$

$$U_{qMt} = \omega_0 M I_{dr} = -R_t I_{qt} - \left(L_t + \frac{1}{\omega_0^2 C_t}\right)\frac{\Delta I_{qt}}{\Delta t} - \left(\omega_0 L_t - \frac{1}{\omega_0 C_t}\right)I_{dt}, \tag{4.45}$$

where U_{dMt} and U_{qMt} are the d-axis and q-axis components of the transmitting coil mutual inductance voltage vector in the dq rotating coordinate system, respectively. Considering the deployment and realization of the actual system, the derivative part mentioned is approximated using a differential formulation. The transmitter coil mutual inductance voltage can then be calculated as:

$$U_{Mt} = \sqrt{U_{dMt}^2 + U_{qMt}^2}$$
$$= \sqrt{\left(-\omega_0 M I_{qr}\right)^2 + \left(\omega_0 M I_{dr}\right)^2} = \omega_0 M I_r. \tag{4.46}$$

The receiving-side equations of the dq dynamic model shown in Eq. (4.30) are analysed. Since the received current cannot be measured directly and I_{dqr} can only be expressed indirectly by Eqs. (4.44) and (4.45), although the change of mutual inductance during the sampling interval time Δt is negligible, the second-order derivatives of I_{dt} and I_{qt} with respect to time (in the form of difference) will be introduced into the calculations of $\Delta I_{dqr}/\Delta t$, which can complicate the estimation process and produce larger errors. Therefore, this term $\Delta I_{dqr}/\Delta t$ should be ignored in the derivation of the method in order to estimate the mutual inductance more directly and quickly. Based on the above analysis, the derivative term in Eq. (4.30) is ignored and multiplied by $\omega_0 M$ on both sides to obtain the following equation:

$$-j\omega_0^2 M^2 I_{dqt} = (R_r + R_L)\omega_0 M I_{dqr} + j\left(\omega_0 L_r - \frac{1}{\omega_0 C_r}\right)\omega_0 M I_{dqr}. \tag{4.47}$$

Solving for the modes on both sides of the above equation yields the following equation for the magnitude information:

$$\omega_0^2 M^2 I_t = \sqrt{(R_r + R_L)^2 + \left(\omega_0 L_r - \frac{1}{\omega_0 C_r}\right)^2} \, \omega_0 M I_r \tag{4.48}$$
$$= |Z_r|\omega_0 M I_r = |Z_r| U_{Mt},$$

where $|Z_r|$ is the known impedance of the receiving side. Since the transmitting coil mutual inductance voltage magnitude U_{Mt} and the transmitting current magnitude I_t can be calculated by Eqs. (4.46) and (4.35), respectively, the following mutual inductance estimation equations are derived on the basis of Eq. (4.48):

$$M = \sqrt{\frac{|Z_r| U_{Mt}}{\omega_0^2 I_t}}. \tag{4.49}$$

Based on the above analytical derivation, Fig. 4.22 illustrates a flowchart of the mutual inductance estimation scheme for the hover charging of UAVs implemented in a digital signal processor. The specific steps of the method are:

1. The system circuit parameters are determined according to the application scenario, including the transmit coil inductance L_t, the transmit-side compensation capacitance C_t, the transmit-side equivalent internal resistance R_t, the system operating angular frequency ω_0, and the receive-side impedance $|Z_r|$.

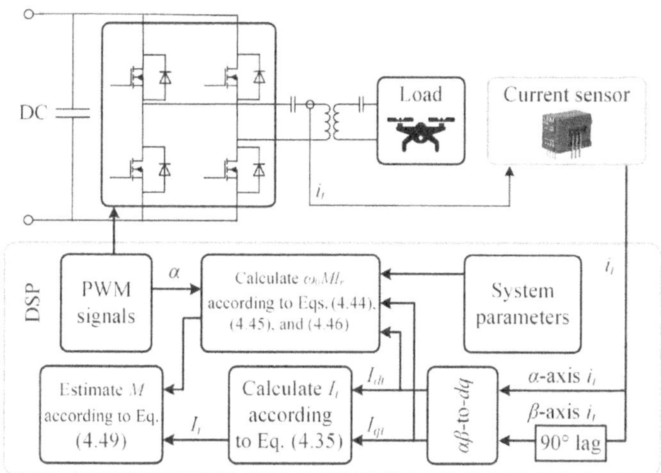

Figure 4.22 Proposed mutual inductance estimation control block diagram.

2. The circuit input voltage U_s is determined based on the phase-shift angle, and the dq transform is completed based on the instantaneous value of the transmit current measured by the sampling, obtaining I_{dt} and I_{qt}.

3. According to the known system circuit parameters – the input voltage amplitude U_s, d-axis component I_{dt}, and q-axis component I_{qt} of the transmit current vector – the mutual inductance voltage amplitude U_{Mt} and transmit current amplitude I_t of the transmit coil are calculated using the dq dynamic model.

4. The system mutual inductance is estimated based on these calculations.

It is easy to see that, when the system circuit parameters are already known, the estimation of mutual inductance can be accomplished through sampling the instantaneous value of the transmit current with the phase-shift control. Except for the system circuit parameters that need to be pre-measured, all the steps in the mutual inductance estimation process are completed on the energy-transmitting side, and only one current sensor needs to be added. Hence, this solution does not add much weight to the UAV side. A major advantage of the scheme is the accelerated estimation speed, which satisfies the demands of UAV hover charging.

4.3.3 Simulation Verification

In this section, MATLAB and Simulink simulation environments are mainly used for analysis and method validation. Table 4.1 describes the simulation parameters. Circuit parameters can be measured from the experimental platform used for method validation with the LCR tester to ensure parameter consistency between the simulation model and the experimental platform. The input voltage amplitude is 7.64 V and mutual inductance is around 10.42 μH. As depicted in Fig. 4.23, the frequency characteristics of the transmit and receive currents, including the amplitude and phase frequencies, under three different load conditions (5 Ω, 10 Ω, and 20 Ω) are obtained from 80 to 120 kHz

Table 4.1 System parameters

Item	Value
Transmitter coil inductance (L_t)	77.016 μH
Transmitter compensation capacitance (C_t)	32.932 nF
Transmitter equivalent resistance (R_t)	0.342 Ω
Receiver coil inductance (L_r)	73.408 μH
Receiver compensation capacitance (C_r)	34.730 nF
Receiver equivalent resistance (R_s)	0.482 Ω
Resistance of load (R_L)	9.92 Ω/19.58 Ω
System operating frequency (f)	100 kHz
DC source voltage (V_{dc})	25 V

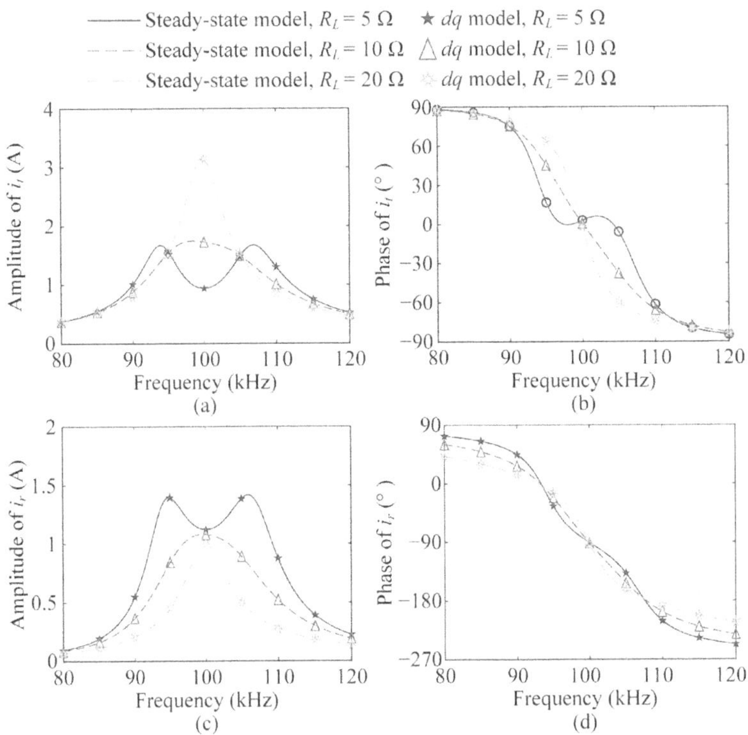

Figure 4.23 Comparative current frequency properties of different models. (a) Current amplitude of i_t. (b) Current phase of i_t. (c) Current amplitude of i_r. (d) Current phase of i_r.

based on the series–series compensated sinusoidal steady-state circuit model of the WPT system. As illustrated in Fig. 4.23, the dq dynamic model is in agreement with the amplitude and frequency property of the common circuit model in these frequency ranges, which confirms the credibility of the constructed dq dynamic model.

Simulations were conducted for various load cases to evaluate the effectiveness of the proposed estimation scheme with the application of the dynamic model. Figure 4.24 illustrates the mutual inductance estimation results for different coupling

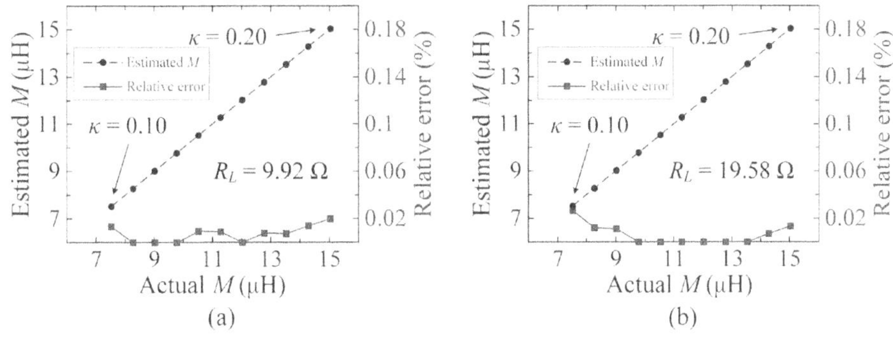

Figure 4.24 Estimated mutual inductance simulation outcome with different loads. (a) $R_L = 9.92\ \Omega$. (b) $R_L = 19.58\ \Omega$.

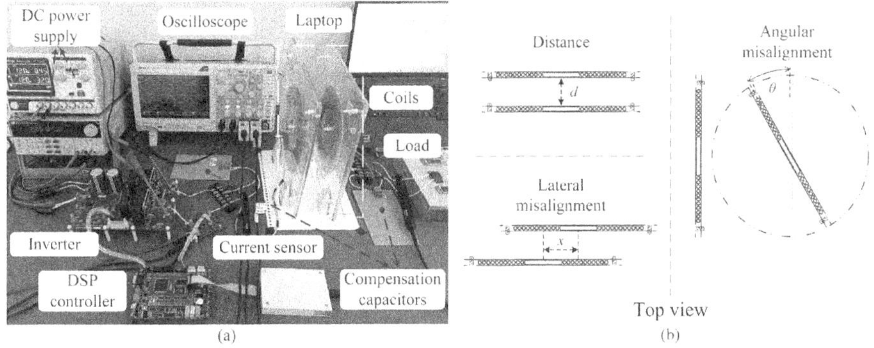

Figure 4.25 (a) Experimental prototype; (b) schematic diagram of the coils.

coefficients ranging from 0.1 to 0.2. This range fully covers the requirements of the hovering charging system. The test results in the figure indicate that the estimation error is less than 0.04% for various load conditions (10 Ω and 20 Ω), which indicates that the proposed mutual inductance estimation method could precisely estimate the mutual inductance with little error, provided that the system parameters and the transmitting current measurements are completely accurate.

4.3.4 Experimental Verification

In order to further evaluate the proposed mutual inductance estimation scheme, an experimental platform for method validation was built, as illustrated in Fig. 4.25. System experimental parameters are as listed in Table 4.1. The system coupling coils are wound with Litz wire to reduce the losses caused by the skin effect and neighbourhood effect at high frequency. Polypropylene film capacitors are used for the compensation capacitors. The loads are high-frequency non-inductive copper plate thick film resistors. The GaN full-bridge inverter powers the WPT system.

The positional relationship between the coils is first clarified as shown in Fig. 4.25(b). Here, d, x, and θ are distance, lateral misalignment, and angular

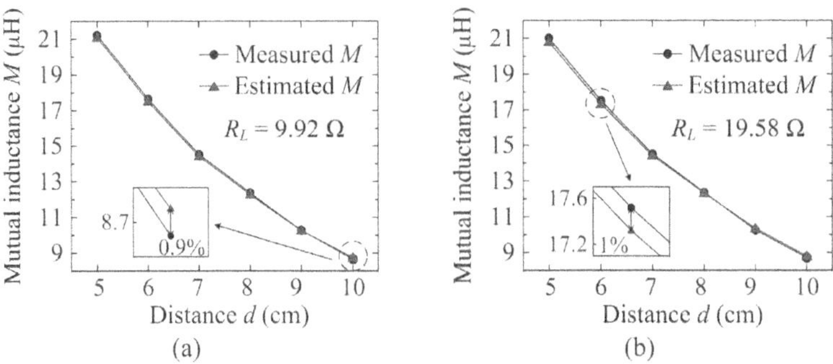

Figure 4.26 Estimated experimental outcome of M with different loads. (a) $R_L = 9.92\ \Omega$; (b) $R_L = 19.58\ \Omega$.

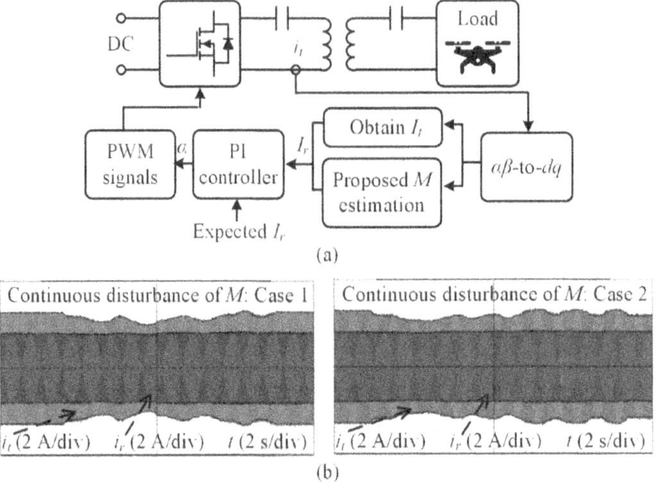

Figure 4.27 (a) A schematic diagram of the control strategy. (b) Test waveforms under continuous position perturbation.

misalignment between the coils, respectively. If x and θ are not explicitly given, the lateral and angular misalignments between the coils are assumed to be 0.

Figure 4.26 illustrates the test results of the actual and predicted values of mutual inductance for different load conditions (for 9.92 Ω and 19.58 Ω) with an air gap of 5–10 cm. As can be seen, the estimated value of mutual inductance is almost equal to the actual value, which demonstrates that the proposed mutual inductance estimation scheme can realize highly accurate mutual inductance estimation in a wide range of situations.

To validate the performance of the estimation scheme for timely tracking of varying coupling coefficients under continuous perturbation of the mutual inductance, an experimental programme for current control (Fig. 4.27(a)) was devised. It should be emphasized that the PI controller aims to achieve constant load current based on accurate mutual inductance prediction. In these experiments, a mutual inductance continuity perturbation in the 7–9 cm air gap was used to simulate the hover charging scenario. Assume that the desired load current is 3.2 A. As shown in Fig. 4.27(b), the receiving current was almost maintained under the mutual inductance continuity perturbation, and the experimental results confirmed that the mutual inductance estimation scheme could rapidly and accurately, in real time, track the changing mutual inductance.

In this section, the novel mutual inductance estimation scheme has been introduced according to the *dq* dynamic model to address the key technical issues that are specific to the hovering wireless charging of UAVs. This aims to track the continuously changing mutual inductance quickly and accurately, in real time, to provide the necessary prerequisites for the realization of stable energy replenishment of the hovering wireless charging system. By using a series–series compensation topology and deploying it on the transmitter side, the lightweight design requirement on the receiver side is met. With the application of the dynamic model, fast tracking of the continuity perturbation of mutual inductance is realized. A combination of simulation and experiment was adopted to evaluate and validate the proposed method in various ways. The test results indicate that the proposed estimation scheme is able to follow the continuously changing mutual inductance successfully, with a relative error of less than 1% and in a time of less than 4 ms. Therefore, this scheme can be effectively applied to the UAV hover charging system to ensure the stable operation of the system.

References

[1] Z. Zhang, H. Pang, A. Georgiadis, and C. Cecati, 'Wireless power transfer: an overview', *IEEE Trans. Ind. Electron.*, vol. 66, no. 2, pp. 1044–1058, Feb. 2019.

[2] Y. Gu, J. Chen, S. Chang, and Z. Zhang, 'Constant power control against M/R with expanded PT-symmetric range for wireless in-flight charging of drones', *IEEE Trans. Magn.*, 2023, doi: 10.1109/TMAG.2023.3284826.

[3] W. Han, K. T. Chau, C. Jiang, W. Liu, and W. H. Lam, 'Design and analysis of quasi-omnidirectional dynamic wireless power transfer for fly-and-charge', *IEEE Trans. Magn.*, vol. 55, no. 7, pp. 1–9, Jul. 2019.

[4] Y. Gu, J. Wang, Z. Liang, Y. Wu, C. Cecati, and Z. Zhang, 'Single-transmitter multiple-pickup wireless power transfer: advantages, challenges, and corresponding technical solutions', *IEEE Ind. Electron. Mag.*, vol. 14, no. 4, pp. 123–135, Dec. 2020.

[5] Y. Gu, J. Wang, Z. Liang, and Z. Zhang, 'Mutual-inductance-dynamic-predicted constant current control of LCC-P compensation network for drone wireless in-flight charging', *IEEE Trans. Ind. Electron.*, vol. 69, no. 12, pp. 12710–12719, Dec. 2022.

[6] J. Zhou, B. Zhang, W. Xiao, D. Qiu, and Y. Chen, 'Nonlinear parity-time-symmetric model for constant efficiency wireless power transfer: application to a drone-in-flight wireless charging platform', *IEEE Trans. Ind. Electron.*, vol. 66, no. 5, pp. 4097–4107, May 2019.

[7] H. Zhang, Y. Chen, C. H. Jo, S. J. Park, and D. H. Kim, 'DC-link and switched capacitor control for varying coupling conditions in inductive power transfer system for unmanned aerial vehicles', *IEEE Trans. Power Electron.*, vol. 36, no. 5, pp. 5108–5120, May 2021.

[8] Z. Zhang, S. Shen, Z. Liang, S. H. K. Eder, and R. Kennel, 'Dynamic-balancing robust current control for wireless drone-in-flight charging', *IEEE Trans. Power Electron.*, vol. 37, no. 3, pp. 3626–3635, Mar. 2022.

[9] K. Chen and Z. Zhang, 'Rotating-coordinate-based mutual inductance estimation for drone in-flight wireless charging systems', *IEEE Trans. Power Electron.*, vol. 38, no. 9, pp. 11685–11693, Sep. 2023.

[10] Y. Gu, J. Wang, Z. Liang, and Z. Zhang, 'A wireless in-flight charging range extended PT-WPT system using S/single-inductor-double-capacitor compensation network for drones', *IEEE Trans. Power Electron.*, vol. 38, no. 10, pp. 11847–11858, Oct. 2023.

[11] Y. Gu, J. Wang, Z. Liang, and Z. Zhang, 'Communication-free power control algorithm for drone wireless in-flight charging under dual-disturbance of mutual inductance and load', *IEEE Trans. Ind. Inform.*, vol. 20, no. 3, pp. 3703–3714, Mar. 2024.

[12] Z.-J. Liao, F. Wu, C.-H. Jiang, Z.-R. Chen, and C.-Y. Xia, 'Analysis and design of ideal transformer-like magnetic coupling wireless power transfer systems', *IEEE Trans. Power Electron.*, vol. 37, no. 12, pp. 15728–15739, Dec. 2022.

[13] C. Cai, S. Wu, L. Jiang, Z. Zhang, and S. Yang, 'A 500-W wireless charging system with lightweight pick-up for unmanned aerial vehicles', *IEEE Trans. Power Electron.*, vol. 35, no. 8, pp. 7721–7724, Aug. 2020.

[14] S. Assawaworrarit, X. Yu, and S. Fan, 'Robust wireless power transfer using a nonlinear parity-time-symmetric circuit', *Nature*, vol. 546, no. 7658, pp. 387–390, Jun. 2017.

[15] Y. Gu, Q. Zhu, L. Jia, G. Li, and Z. Zhang, 'Optimized high-order compensation topology for PT-WPT systems with expanded constant power region', *IEEE Trans. Magn.*, 2023, doi: 10.1109/TMAG.2023.3287481.

[16] F. Liu, Y. Yang, Z. Ding, X. Chen, and R. Kennel, 'A multifrequency superposition methodology to achieve high efficiency and targeted power distribution for a multiload MCR WPT system', *IEEE Trans. Power Electron.*, vol. 33, no. 10, pp. 9005–9016, 2018.

[17] X. Dai, X. Li, Y. Li, and A. P. Hu, 'Maximum efficiency tracking for wireless power transfer systems with dynamic coupling coefficient estimation', *IEEE Trans. Power Electron.*, vol. 33, no. 6, pp. 5005–5015, Jun. 2018.

[18] J. Hu, J. Zhao, and C. Cui, 'A wide charging range wireless power transfer control system with harmonic current to estimate the coupling coefficient', *IEEE Trans. Power Electron.*, vol. 36, no. 5, pp. 5082–5094, May 2021.

[19] Y. Liu and H. Feng, 'Maximum efficiency tracking control method for WPT system based on dynamic coupling coefficient identification and impedance matching network', *IEEE J. Emerg. Sel. Top. Power Electron.*, vol. 8, no. 4, pp. 3633–3643, Dec. 2020.

[20] W. Li, G. Wei, C. Cui, X. Zhang, and Q. Zhang, 'A double-side self-tuning LCC/S system using a variable switched capacitor based on parameter recognition', *IEEE Trans. Ind. Electron.*, vol. 68, no. 4, pp. 3069–3078, Apr. 2021.

[21] J. Yin, D. Lin, T. Parisini, and S. Y. R. Hui, 'Front-end monitoring of the mutual inductance and load resistance in a series–series compensated wireless power transfer system', *IEEE Trans. Power Electron.*, vol. 31, no. 10, pp. 7339–7352, Oct. 2016.

[22] J. Zhou, B. Zhang, W. Xiao, D. Qiu, and Y. Chen, 'Nonlinear parity-time-symmetric model for constant efficiency wireless power transfer: application to a drone-in-flight wireless charging platform', *IEEE Trans. Ind. Electron.*, vol. 66, no. 5, pp. 4097–4107, May 2019.

[23] W. Liu, K. T. Chau, C. H. T. Lee, X. Tian, and C. Jiang, 'Hybrid frequency pacing for high-order transformed wireless power transfer', *IEEE Trans. Power Electron.*, vol. 36, no. 1, pp. 1157–1170, Jan. 2021.

[24] T. Kan, F. Lu, T. Nguyen, P. P. Mercier, and C. C. Mi, 'Integrated coil design for EV wireless charging systems using LCC compensation topology', *IEEE Trans. Power Electron.*, vol. 33, no. 11, pp. 9231–9241, Nov. 2018.

[25] Z. Zhang, K. T. Chau, C. Qiu, and T. W. Ching, 'A positioning-tolerant wireless charging system for roadway-powered electric vehicles', *J. Appl. Phys.*, vol. 117, no. 17, pp. 1–4, Mar. 2015.

[26] Z. Zhang, H. Pang, C. H. T. Lee, X. Xu, X. Wei, and J. Wang, 'Comparative analysis and optimization of dynamic charging coils for roadway-powered electric vehicles', *IEEE Trans. Magn.*, vol. 53, no. 11, pp. 1–6, Nov. 2017.

[27] H. Hao, G. A. Covic, and J. T. Boys, 'A parallel topology for inductive power transfer power supplies', *IEEE Trans. Power Electron.*, vol. 29, no. 3, pp. 1140–1151, Mar. 2014.

5   Control Scheme based on the PT-Symmetric WPT System

5.1   Introduction

Parity-time (PT) symmetry originated from quantum physics and has been introduced to WPT systems in recent years [1, 2]. It has the great merit of high efficiency while maintaining a constant output against changing coupling strength. Distinguishing it from the common magnetically coupled WPT system, this new PT-WPT system can use the DC power source and the inverter as a kind of negative resistance $-R_N$. Compared with other control strategies [3–7], the PT-WPT system allows automatic frequency regulation over a certain area (i.e. the PT region) to maintain a constant output. The control technology of the system is required only on the transmitter side and there are no additional communication units or extra converters. This helps to increase the real-time capability of the system and reduces the weight on the receiving side. These benefits are particularly useful for UAV hover charging.

However, when battery load varies during the charging process, the load charging power of this PT-WPT system still changes with the variable load, which needs to be overcome. The constant power in cases of mutual induction disturbance cannot be flexibly regulated, which is a limitation in some application scenarios with variable charging needs. Hence, this chapter first introduces the models and analysis of the PT-symmetric WPT system on the basis of the circuit theory and coupled-mode theory. In order to solve the multiple problems of load and coupling coefficients in practical applications, a charging power regulation strategy under the PT-WPT system is presented on the basis of dynamic load identification.

For the PT-WPT system, an important condition for realizing constant power (CP) charging is that the coupling strength is above a fixed level. That is to say, under the weak coupling condition of the WPT system (i.e. a more distanced air gap) it is difficult for the system charging power to overcome variations in coupling strength. To address this key issue we need to realize the expansion of the stable charging range of the PT-WPT system as well as flexible adjustment to meet the different demands of varied application scenarios. In order to solve this problem, two practical methods of CP region expansion and adjustment are presented in Chapter 6, where the novel high-order compensation network in [8] is introduced and experimentally verified. The flexible charging range extension method is presented with the primary-side modulation scheme, which shows the salient features of convenience and high system flexibility.

5.2   Models and Analysis for the PT-Symmetric WPT System

There are two theory models of analysis for the PT-WPT system: circuit theory (CT) and coupled-mode theory (CMT). Compared with the CT model, the CMT model can simplify the analysis by reducing the order of the differential equations by half [9]. The CMT model is particularly suitable for higher-order topologies and is useful for describing the resonant coupling units in a WPT system from the perspective of energy transfer. In the following, the series–series compensation network of the PT-WPT system will be modelled and analysed by the CT model and the CMT model.

5.2.1   Modelling based on Circuit Theory

The circuit model is a method to build an electrical model of a system on the basis of Kirchhoff's law and mutual inductance theory, so as to analyse the internal parameters of the system. It is suitable for analysing and optimizing the electrical parameters of the system and comparing the transmission characteristics of different topologies.

The system circuit block diagram for UAV hovering wireless charging on the basis of the PT-symmetric energy mirroring mechanism is shown in Fig. 5.1. It mainly consists of an inverter, a transmitting-side circuit, a receiving-side circuit, a rectifier bridge, and a battery load. The inverter of the WPT system is composed of four switching tubes that generate the AC voltage. The transmitting and receiving circuits are coupled together by two coupling coils, and both sides have a compensation topology to ensure the wireless transmission of energy (generally a series–series type compensation topology is applied). In the receiving-side circuit, the induced voltage is compensated and connected to a rectifier bridge and a filter circuit to transform the AC voltage into a DC voltage of a certain amplitude in order to power the drone. To satisfy the application demands of UAV hovering wireless charging, the series–series topology is more advantageous due to its simple structure, low weight, and stable output. As shown in Fig. 5.1, S1–S_4 constitute a GaN inverter that converts the DC voltage to an AC voltage to provide energy for the whole WPT system. In the adopted

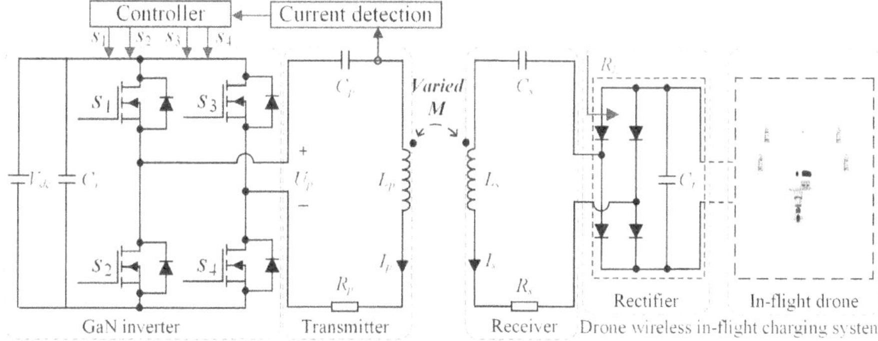

Figure 5.1 Hover charging system schematic.

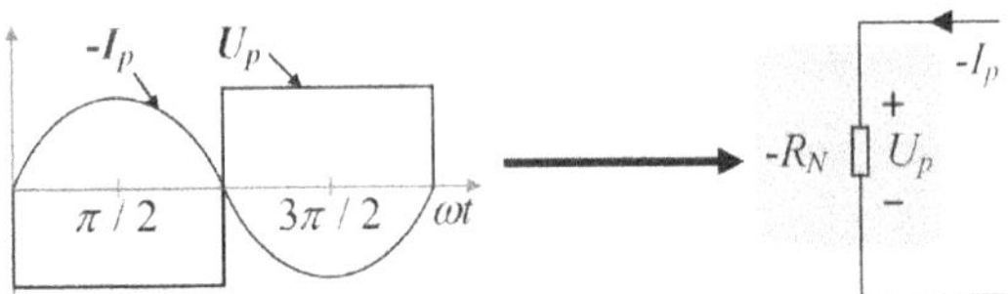

Figure 5.2 System negative resistance.

topological network, the transmitter and receiver sides include a coil, series capacitors, and equivalent loads (L_p, C_p, R_p and L_s, C_s, R_s), respectively. The frequent fluctuation of the position of the drone in the hovering state causes the mutual inductance M of the coupling coils to vary. This rectified and filtered drone battery R_{bat} can be equated to the load R_L on the output side. Here, we can be satisfied that $R_L = 8R_{bat}/\pi^2$.

One of the features of this PT-WPT system is that this closed-loop control by the primary-side circuit design ensures that this inverter output voltage will be 180 degrees out of phase with the transmitting current, as depicted in Fig. 5.2. After the later theoretical analysis, it can be seen that this PT-WPT system can be ensured by implementing closed-loop regulation. Hence, stable output characteristics – including load power and efficiency – can be achieved in a given region of this system [10–13], which is different from the conventional magnetic coupling resonant (MCR) WPT system [14, 15]. Its operating frequency can be adaptively varied to accommodate mutual inductance variation.

By modelling and analysing the PT-WPT system illustrated in Fig. 5.3(a), the equivalent circuit model of this system is expressed:

$$\begin{bmatrix} \left(-R_N + R_p\right) + j\omega L_p + \dfrac{1}{j\omega C_p} & j\omega M \\ j\omega M & (R_s + R_L) + j\omega L_s + \dfrac{1}{j\omega C_s} \end{bmatrix} \begin{bmatrix} I_p \\ I_s \end{bmatrix} = 0, \qquad (5.1)$$

where the system switching frequency is defined as ω, and I_p and I_s are the primary and pickup currents, respectively. This system requires that the primary side and the pickup side satisfy the same natural resonant frequency, which is:

$$\omega_0 = 1/\sqrt{L_p C_p} = 1/\sqrt{L_s C_s}. \qquad (5.2)$$

When the above resonance conditions are satisfied, Fig. 5.3(b) illustrates this phasor diagrams of this system's state variables. Then, Eq. (5.1) can be expressed as:

$$\begin{bmatrix} -2g_1 + j\left(\omega - \omega_0^2/\omega\right) & j\omega\kappa\sqrt{L_s/L_p} \\ j\omega k\sqrt{L_p/L_s} & 2\gamma_2 + j\left(\omega - \omega_0^2/\omega\right) \end{bmatrix} \begin{bmatrix} I_p \\ I_s \end{bmatrix} = 0. \qquad (5.3)$$

In the above equations, the gain ratio at the primary side and the loss ratio at the receiver side of the system are expressed by g_1 and γ_2, which satisfy $g_1 = (R_N - R_p)/2L_p$ and $\gamma_2 = (R_L + R_s)/2L_s$, respectively. This coupling coefficient of the coupling

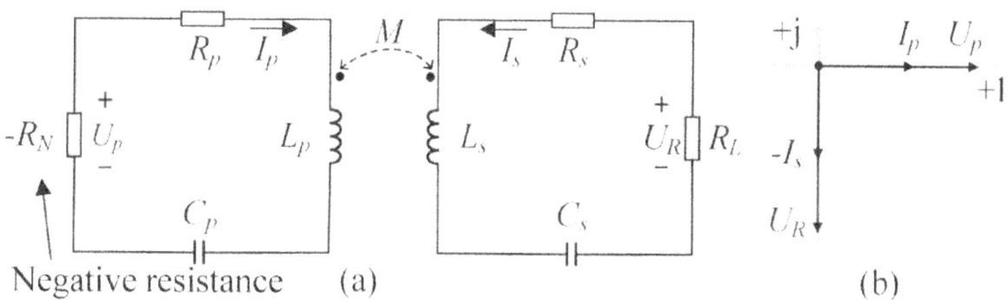

Figure 5.3 (a) System equivalent circuit diagram; (b) phasor diagram.

coils can be expressed as $\kappa = M/\sqrt{L_p L_s}$. Based on the above equations, the characteristic equation about the operating angular frequency ω is given as:

$$\left[-2g_1 + j\left(\omega - \omega_0^2/\omega\right)\right]\left[2\gamma_2 + j\left(\omega - \omega_0^2/\omega\right)\right] + \omega^2\kappa^2 = 0. \tag{5.4}$$

After separating the imaginary and real components of this characteristic equation, it is derived as Eq. (5.5). At the same time, the steady-state solution for the operating angular frequency of the system is obtained as Eq. (5.6):

$$\begin{aligned} (2\gamma_2 - 2g_1)\left(\omega - \omega_0^2/\omega\right) &= 0, \\ -4g_1\gamma_2 - \left(\omega - \omega_0^2/\omega\right)^2 + \omega^2\kappa^2 &= 0. \end{aligned} \tag{5.5}$$

$$\omega = \begin{cases} \omega_{1,2} = \sqrt{\dfrac{\omega_0^2 - 2\gamma_2^2 \pm \sqrt{\left(\omega_0^2 - 2\gamma_2^2\right)^2 - \omega_0^4(1 - \kappa^2)}}{1 - \kappa^2}}, \kappa \geq \kappa_c. \\ \omega_0, \kappa < \kappa_c \end{cases} \tag{5.6}$$

Here, the system critical coupling coefficient is defined as κ_c and is given as:

$$\kappa_c = \sqrt{1 - \left[1 - \frac{(R_L + R_s)^2}{2\omega_0^2 L_s^2}\right]^2}. \tag{5.7}$$

Based on Eq. (5.7), this PT-WPT system could be divided into two regions of stable operation: the exact PT symmetry range ($\kappa \geq \kappa_c$) and the broken PT symmetry range ($\kappa < \kappa_c$).

Exact PT Symmetry Range

In the strong coupling region, this PT-WPT system enters this exact PT symmetry range when $\kappa \geq \kappa_c$. At this time, this gain g_1 and loss γ_2 are equal and satisfy $(R_N - R_p)/L_p = (R_L + R_s)/L_s$. According to Eq. (5.6), the system operating angular frequency is stabilized at $\omega_{1,2}$. Figure 5.4 illustrates the system equivalent circuit diagrams in this exact PT symmetry region for the different sides. It can be seen that the state variables of the system are not affected by mutual inductance. According to Eqs. (5.3) and (5.6), the relationship between the transmitting-side and the pickup-side currents could be obtained as follows:

$$I_p/I_s = \sqrt{L_s/L_p}. \tag{5.8}$$

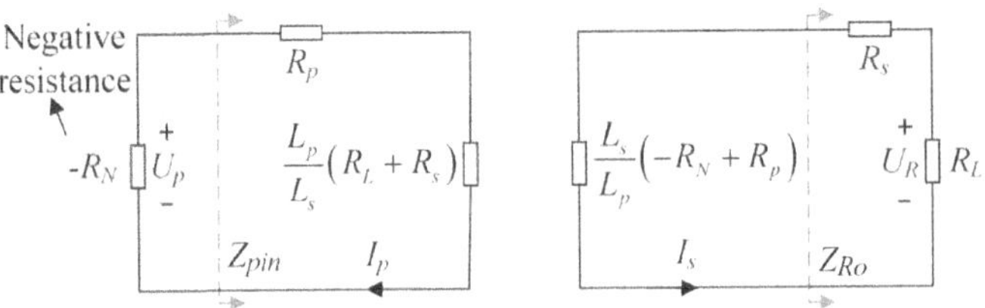

Figure 5.4 Equivalent circuit of a system with reflected impedance.

System load power P_{out} and transmission efficiency η could be given as:

$$P_{out} = I_s^2 R_L = U_p^2 R_L \bigg/ \left[\sqrt{\frac{L_s}{L_p}} R_p + \sqrt{\frac{L_p}{L_s}} (R_L + R_s) \right]^2, \tag{5.9}$$

$$\eta = \frac{I_s^2 R_L}{I_p^2 R_N} = \frac{R_L}{R_L + R_s + R_p L_s / L_p}. \tag{5.10}$$

Based on the above equations, it can be found that, for the circuit model, system load power and transmission efficiency do not suffer from coupling coefficient κ at this exact PT symmetry range.

Broken PT Symmetry Range

Under this weak coupling condition, when $\kappa < \kappa_c$, the PT-WPT system enters the broken PT symmetry range, and the system work frequency is maintained at the natural resonant frequency ω_0, where Eq. (5.11) can be met:

$$-R_N = \frac{\omega^2 M^2}{R_L + R_s} - R_p \tag{5.11}$$

According to Eqs. (5.3) and (5.11), the system load power and transfer efficiency are derived as:

$$P_{out} = I_s^2 R_L = U_p^2 R_L \omega^2 M^2 \big/ \left[\omega^2 M^2 + R_p (R_L + R_s) \right]^2, \tag{5.12}$$

$$\eta = \frac{I_s^2 R_L}{I_p^2 R_N} = \frac{\omega^2 M^2 R_L}{\omega^2 M^2 (R_L + R_s) + R_p (R_L + R_s)^2}. \tag{5.13}$$

From these equations, it can be found that the load power and transfer efficiency of this PT-WPT system can still be affected by the coupling coefficients in this broken state, and therefore this region is undesirable. In order to enlarge the CP charging range of this PT-WPT system, it is necessary to reduce the coupling coefficient κ_c, which in turn meets the demands of our application scenarios.

5.2.2 Modelling Based on Coupled-Mode Theory

The CMT model can also be used to model and analyse the PT-WPT system. It has the advantage of being suitable for analysing the energy transfer process of the coupled resonators of the system. The CMT analyses the coupled perturbations of the charging resonators with the help of the study of coupled oscillations, which visually describes the process of the energy transfer of this WPT system, and is a kind of approximation of the modelling method. Compared with the circuit model, the CMT model simplifies the analysis by reducing the order of the differential equations by half [16]. The system circuit equation can be represented as:

$$L_p \frac{di_p}{dt} + M_{ps} \frac{di_s}{dt} + u_{cp} + i_p R_p = u_p,$$

$$M_{ps} \frac{di_p}{dt} + L_s \frac{di_s}{dt} + u_{cs} + i_s(R_s + R_{eq}) = 0,$$

$$C_p \frac{du_{cp}}{dt} = i_p,$$

$$C_s \frac{du_{cs}}{dt} = i_s.$$

(5.14)

According to the fundamental harmonic approximation, the relationship of this fundamental component of the inverter output voltage and DC input voltage satisfies $u_p = \frac{4V_{dc}}{\sqrt{2}\pi}$. Based on this CMT model, system energy modes ψ_p and ψ_s of resonators on the transmitter and receiver sides could be expressed as:

$$\psi_p = \Psi_p e^{j(\omega t + \theta_p)} = \sqrt{\frac{L_p}{2}} i_p + j\sqrt{\frac{C_p}{2}} u_{cp},$$

$$\psi_s = \Psi_s e^{j(\omega t + \theta_s)} = \sqrt{\frac{L_s}{2}} i_s + j\sqrt{\frac{C_s}{2}} u_{cs}.$$

(5.15)

In these equations, the system operating angular frequency is denoted by ω. The magnitude and phase of the energy modes on the primary and pickup sides can be denoted with Ψ_p, Ψ_s and θ_p, θ_s, respectively. According to Eqs. (5.14) and (5.15), the relationship between the system state variable and this energy mode could be obtained as:

$$i_p = \Psi_p \sqrt{\frac{2}{L_p}} \cos(\omega t + \theta_p),$$

$$u_{cp} = \Psi_p \sqrt{\frac{2}{C_p}} \sin(\omega t + \theta_p),$$

$$i_s = \Psi_s \sqrt{\frac{2}{L_s}} \cos(\omega t + \theta_s),$$

$$u_{cs} = \Psi_s \sqrt{\frac{2}{C_s}} \sin(\omega t + \theta_s).$$

(5.16)

In this PT-WPT system, the natural resonant angular frequencies ω_p and ω_s of the transmitting and receiving resonators should match the following equations: $\omega_p = \omega_s = \omega_0$. Then, by substituting Eq. (5.16) into Eq. (5.14), the system dynamics equations based on this CMT model could be deduced as Eq. (5.17) through this averaging scheme [17]. By substituting Eq. (5.17) into Eq. (5.16) and neglecting the high-order parts, the dynamical expression of the system could be derived as:

$$
\frac{d\Psi_p}{dt} = \frac{1}{1-\kappa^2}\left[\frac{2V_{dc}}{\pi\sqrt{2L_p}} - \frac{R_p}{2L_p}\Psi_p + \frac{\kappa(R_s + R_{eq})}{2L_s}\right.
$$

$$
\left. \times \Psi_s \cos(\theta_p - \theta_s) - \frac{\kappa\omega_s}{2}\Psi_s \sin(\theta_p - \theta_s)\right],
$$

$$
\Psi_p\left(\omega + \frac{d\theta_p}{dt}\right) = \frac{\omega_p}{2}\Psi_p + \frac{1}{1-\kappa^2}\left[\frac{\omega_p}{2}\Psi_p - \frac{\kappa\omega_s}{2}\right.
$$

$$
\left. \times \Psi_s \cos(\theta_p - \theta_s) - \frac{\kappa(R_s + R_{eq})}{2L_s}\Psi_s(\theta_p - \theta_s)\right],
$$

$$
\frac{d\Psi_s}{dt} = \frac{1}{1-\kappa^2}\left[-\frac{2\kappa V_{dc}}{\pi\sqrt{2L_p}}\cos(\theta_p - \theta_s) - \frac{R_s + R_{eq}}{2L_s}\Psi_s\right.
$$

$$
\left. + \frac{\kappa R_p}{2L_p}\Psi_p \cos(\theta_p - \theta_s) + \frac{\kappa\omega_p}{2}\Psi_p \sin(\theta_p - \theta_s)\right],
$$

$$
\Psi_s\left(\omega + \frac{d\theta_s}{dt}\right) = \frac{\omega_s}{2}\Psi_s + \frac{1}{1-\kappa^2}\left[-\frac{2\kappa V_{dc}}{\pi\sqrt{2L_p}}\sin(\theta_p - \theta_s)\right.
$$

$$
\left. + \frac{\omega_s}{2}\Psi_s - \frac{\kappa\omega_p}{2}\Psi_p \cos(\theta_p - \theta_s) + \frac{\kappa R_p}{2L_p}\Psi_p \sin(\theta_p - \theta_s)\right],
$$

$$
(5.17)
$$

$$
\frac{d}{dt}\begin{bmatrix}\psi_p \\ \psi_s\end{bmatrix} = \begin{bmatrix} j\omega_0 + \dfrac{1}{\Psi_p}\dfrac{2V_{dc}}{\pi\sqrt{2L_p}} - \dfrac{R_p}{2L_p} & -j\kappa\dfrac{\omega_0}{2} \\[2ex] -j\kappa\dfrac{\omega_0}{2} & j\omega_0 - \dfrac{R_s + R_{eq}}{2L_s} \end{bmatrix}\begin{bmatrix}\psi_p \\ \psi_s\end{bmatrix}, \qquad (5.18)
$$

where the system coupling coefficient could be expressed as: $\kappa = M/\sqrt{L_p L_s}$. Based on Eq. (5.18), the system frequency property expression could be expressed as:

$$
\begin{bmatrix} j(\omega_0 - \omega) + \dfrac{1}{\Psi_p}\dfrac{2V_{dc}}{\pi\sqrt{2L_p}} - \dfrac{R_p}{2L_p} & -j\kappa\dfrac{\omega_0}{2} \\[2ex] -j\kappa\dfrac{\omega_0}{2} & j(\omega_0 - \omega) - \dfrac{R_s + R_{eq}}{2L_s} \end{bmatrix} = 0. \qquad (5.19)
$$

By dividing the real and imaginary components of Eq. (5.19), it can be represented as Eq. (5.20), and the stationary value of the working frequency ω can be obtained as Eq. (5.21):

$$
\frac{\kappa^2\omega_0^2}{4} - (\omega - \omega_0)^2 + \left(\frac{R_p}{2L_p} - \frac{1}{\Psi_p}\frac{2V_{dc}}{\pi\sqrt{2L_p}}\right)\left(\frac{R_s + R_{eq}}{2L_s}\right) = 0,
$$

$$
(\omega - \omega_0)\left(\frac{1}{\Psi_p}\frac{2V_{dc}}{\pi\sqrt{2L_p}} - \frac{R_p}{2L_p} - \frac{R_s + R_{eq}}{2L_s}\right) = 0,
$$

$$
(5.20)
$$

$$\omega = \begin{cases} \omega_0 \pm \sqrt{\left(\frac{\kappa\omega_0}{2}\right)^2 - \left(\frac{R_s+R_{eq}}{2L_s}\right)^2} = \omega_{1,2}, \kappa \geq \kappa_c \\ \omega_0, \kappa < \kappa_c \end{cases} \tag{5.21}$$

In Eq. (5.21), the system critical coupling coefficient is denoted as κ_c:

$$\kappa_c = \frac{R_s + R_{eq}}{\omega_0 L_s}. \tag{5.22}$$

From the above expression, as can be observed, if ω_0 is given, the value of κ_c is only related to the parameters of the receiving side of the WPT system. As the receiver-side inductor L_s increases or the equivalent load R_{eq} decreases (the internal resistor R_s is small relative to R_{eq} and can be neglected), the value of κ_c decreases. Therefore, according to Eq. (5.22), the PT-WPT system has two regions of stable operation: this exact PT symmetry range ($\kappa \geq \kappa_c$) and this broken PT symmetry range ($\kappa < \kappa_c$).

Exact PT Symmetry Region ($\kappa \geq \kappa_c$)

In this region of exact PT symmetry, according to Eq. (5.21), this system's working angular frequency can be automatically stabilized at $\omega_{1,2}$ instead of ω_0, and will vary with the coupling coefficient κ. According to Eqs. (5.20) and (5.21), the amplitude of the energy modes can be expressed as Eq. (5.23).

$$\Psi_p = \Psi_s = \frac{2\sqrt{2L_p}L_s V_{dc}}{\pi\left[R_p L_s + \left(R_s + R_{eq}\right)L_p\right]}. \tag{5.23}$$

The system load power P_{out} and system efficiency η are expressed as:

$$P_{out} = \frac{R_{eq}\Psi_s^2}{L_s} = \frac{8V_{dc}^2 R_{eq}L_p L_s}{\pi^2\left[R_p L_s + \left(R_s + R_{eq}\right)L_p\right]^2}, \tag{5.24}$$

$$\eta = \frac{R_o\Psi_s^2/L_s}{R_p\Psi_p^2/L_p + \left(R_s + R_{eq}\right)\Psi_s^2/L_s} = \frac{R_{eq}}{R_p\left(L_s/L_p\right) + R_s + R_{eq}}. \tag{5.25}$$

From these system output properties, it can be observed that at this exact PT symmetry range, even if mutual inductance changes, the system load power and transmission efficiency are not affected, which is far superior for UAV hover charging. However, it is worth noting that this CP region will only be maintained if κ is greater than κ_c, and κ_c is predominantly related to the parameters R_{eq} and L_s on the receiving side. The smaller the value of κ_c, the larger the constant power region is. Therefore, in order to obtain a large range for the constant power charging region for a UAV hover charging system, it is necessary to reduce κ_c to this desired smaller value.

Broken PT Symmetry Region ($\kappa < \kappa_c$)

At this broken PT range, this system's working angular frequency will be stabilized at ω_0. The system load power and efficiency are calculated as Eqs. (5.26) and (5.27), which are identical to those of the conventional no-feedback MCR-WPT system:

$$P_{out} = \frac{8V_{dc}^2 R_{eq}\omega_0^2\kappa^2 L_p L_s}{\pi^2\left[\omega_0^2\kappa^2 L_p L_s + R_p\left(R_s + R_{eq}\right)\right]^2}, \tag{5.26}$$

$$\eta = \frac{R_{eq}\omega_0^2\kappa^2 L_p L_s}{\omega_0^2\kappa^2 L_p L_s\left(R_s + R_{eq}\right) + R_p\left(R_s + R_{eq}\right)^2}. \tag{5.27}$$

Therefore, under this broken range, system output characteristics will still be affected by the mutual inductance change. To obtain robust system charging characteristics, it should be ensured that the system coupling strength is higher than a specific value (namely κ_c in Eq. (5.22)).

5.2.3 Output Characteristic

Based on the above theoretical derivation, the output characteristics of this PT-WPT system could be concluded regardless of the CMT or CT model – that is, at this exact PT symmetry range, system load power and transmission efficiency will be constant under mutual inductance variation. In the following, these system output properties will be analysed on the basis of the CMT model. First, the similarities and differences between the PT-WPT system and the conventional no-feedback MCR-WPT system are compared and analysed, highlighting the strengths of the PT-WPT system. Then, key output characteristics of the PT-WPT system – the PT-symmetric state range and its influencing factors on the output power – are analysed. The influence of the system variables on the value of κ_c and the effect of the laws of system load to load power are obtained, and provide the theoretical basis for the decreasing scheme of κ_c and load power regulation in the following chapter.

First, we compare the similarities and differences between the PT-WPT system and this common no-feedback MCR-WPT system. As mentioned earlier, the voltage source and inverter of the PT-WPT system could be equated to a negative resistance. In fact, for a fully resonant MCR-WPT system, the voltage source can also be considered as a negative resistance; that is to say, the inverter output voltage u_p and current $-i_p$ across the equivalent negative resistance are also inverted. However, the working frequency of this no-feedback MCR-WPT system will be fixed and there is no feedback circuit to adjust it. For the conventional PT-WPT system, this control circuit design of the negative resistor is the key that provides the possibility of automatic frequency variation. Accordingly, the frequency of this PT-WPT system varies with mutual inductance variation, and thus the expected constant charging power and efficiency can be obtained. Table 5.1 compares these two systems more clearly.

To show the superiority of the PT-WPT system, the load power P_{out} and efficiency η of this PT-WPT system can also be contrasted with those of the common no-feedback MCR-WPT system. Using the experimental parameters (discussed later), the output properties of the two WPT systems are illustrated in Fig. 5.5. It can be observed that the load power of the PT-WPT system tends to be higher when $\kappa \geq \kappa_c$.

Table 5.1 Comparison of the PT-WPT system with the conventional no-feedback MCR-WPT system

	PT-WPT system	**MCR-WPT system**
Control scheme	Current feedback required	No feedback
Work frequency	Adaptive frequency adjustment	Fixed frequency
Output power	Constant (in exact PT range)	Variable
Efficiency	Constant (in exact PT range)	Variable

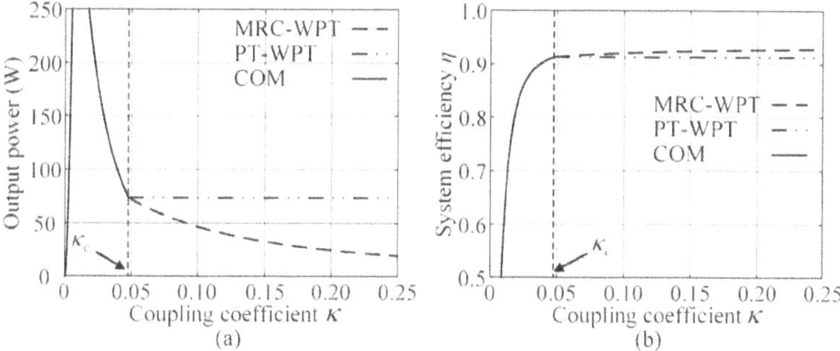

Figure 5.5 Comparison of system output properties: (a) load power (b) efficiency.

In particular, the larger κ is, the more obvious the advantage (higher power) of the PT-WPT system is. As illustrated in Fig. 5.5(b), the transfer efficiency of the PT-WPT system is a little lower than that of the MCR-WPT system. If $\kappa < \kappa_c$, the output properties of the two systems are identical, as illustrated in Fig. 5.5. It is especially noted that in the MCR-WPT system, the output properties will vary greatly as the air gap between the two coupling coils changes, which should be avoided for the safety of the batteries and the power-using equipment.

In the PT-WPT system, the robust load power and efficiency will be guaranteed against κ if $\kappa \geq \kappa_c$. Therefore, it can be concluded that in the strongly coupled state, the PT-WPT system has higher load power without significant efficiency degradation. Also, robust charging behaviour will be guaranteed in the case of coupling strength perturbation. Since the PT-WPT system belongs to the required original edge control strategy, it becomes an attractive alternative to conquer the continuous variation in charging coupling strength of a hovering UAV by virtue of its high robustness and real-time performance.

In order to meet the demand for stable charging power in a large range area in different application scenarios, it is necessary to expand the symmetry region of this PT-WPT system. According to the above theory analyses, the value of κ_c should be reduced to obtain a large range in the constant power region. According to Eq. (5.22), the value of κ_c is mainly related to the equivalent load R_{eq}, the intrinsic resonance angular frequency ω_0, and the receiving-side inductor L_s. Theoretically, κ_c will reduce with a reduction of the equivalent load R_{eq}, an increase of the receiver coil L_s, and an increase of the operating angular frequency $\omega_0(f)$, which in turn expands the symmetry range of this PT-WPT system.

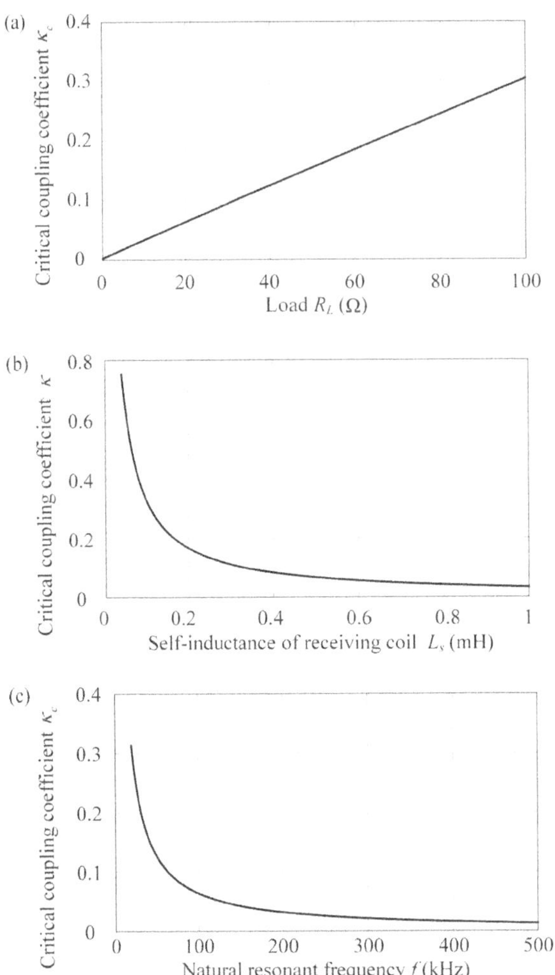

Figure 5.6 The relationship between critical coupling coefficients and system parameters: (a) load; (b) self-inductance of the receiving coil; (c) natural resonant frequency.

Figure 5.6 depicts the relationship of the system parameters to the critical coupling coefficient. From Fig. 5.6(a), it can be observed that as the equivalent load R_{eq} decreases, the critical coupling coefficient κ_c also decreases, which expands the symmetry region of the system. However, according to Eq. (5.25) it can be found that the system efficiency decreases as the equivalent load R_{eq} decreases. In wireless charging systems the equivalent load R_{eq} is generally determined based on the application requirements and cannot be changed or reduced arbitrarily. In addition, by means of added power converters, it is possible to get a smaller equivalent load R_{eq}, thereby extending this robust system charging range. However, these methods can usually only be implemented on the system pickup side, which increases the hovering drone payload and results in more power loss. Therefore, expanding the symmetry region through decreasing the equivalent load of the WPT system would be limited by practical applications.

As can be observed from Fig. 5.6(b), the critical coupling coefficient decreases following the increase of the receiving coil L_s, which expands the symmetry region of the system. The value of κ_c is reduced by increasing the pickup-side inductor, and so is greater when the pickup-side inductor is small. However, coil self-inductance is generally proportional to its volume and weight (i.e. if the value of L_s increases, its volume and weight will increase accordingly). In wireless charging systems (especially for UAV in-flight charging), the space and weight allowed for the receiving coil should be as small as possible. Thus, L_s has a maximum value, and once the maximum value is reached the constant power region cannot be further expanded. Therefore, the limited payload capacity and space constraints of drones impose stringent requirements for the receiving device of the WPT system, and simply increasing the receiving coil L_s is a limited option.

As can be observed from Fig. 5.6(c), the value of κ_c reduces with the increase of the natural resonance angular frequency ω_0, which expands the system PT symmetry region. However, for some special applications of wireless charging there is a range of operating frequencies required, so ω_0 cannot be increased infinitely to expand the symmetry region. At the same time, the increase in operating frequency presents a challenge to the inverter or rectifier bridge switching devices in the WPT system, which requires the switching devices to withstand higher frequencies. Moreover, the high frequency increases the power loss of the switching devices.

In summary, the constant power range of the PT-WPT system under the conventional network is limited, and the exact PT symmetry range needs to be expanded and adjusted – that is, a larger transfer distance is required to satisfy the application requirements in various cases. The symmetry region (i.e. κ_c) of the PT-WPT system is mainly related to system load R_{eq}, natural resonant angular frequency ω_0, and the receiving-side inductor L_s (i.e. the extension and adaptive adjustment of the symmetry range could be achieved with the regulation of these three parameters). On the basis of the above theory analyses, this mechanism and scheme for expanding the symmetry region will be investigated in the rest of this chapter. A high-order topology and a primary-side modulation strategy are introduced in the next chapter to achieve the reduction in the equivalent load of the system, which in turn expands the system symmetry range of the PT-WPT system.

According to the above theoretical analysis of the PT-WPT system, the main benefit is the achievement of robust output under disturbed mutual inductance at this PT symmetry range – that is, mutual inductance decoupling is achieved. At this PT symmetry range, a robust output is achieved by the adaptive adjustment of the system's working frequency. Figure 5.7 shows the system output properties against various loads in this region. As can be observed, even in this PT symmetrical region the output properties remain variable with perturbations of load R_{eq}. The system load power first increases and then decreases with increasing R_{eq}, while the transfer efficiency will continue to rise as load increases. There exists a maximum output power value of $U_p^2/(4R_sL_p/L_s + 4R_p)$ where $R_{eq} = R_s + R_pL_s/L_p$. To observe in a more

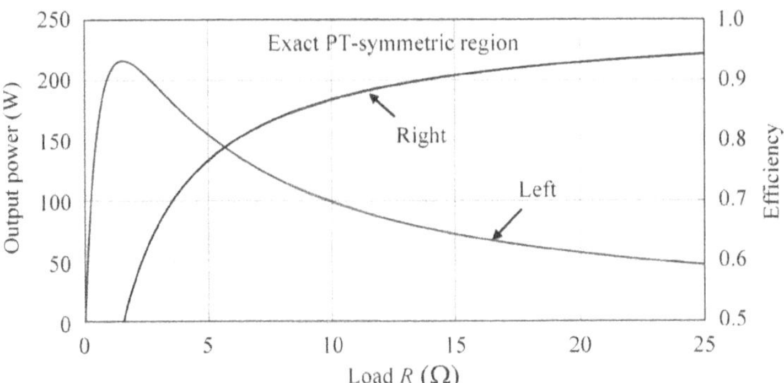

Figure 5.7 System load power and efficiency against loads.

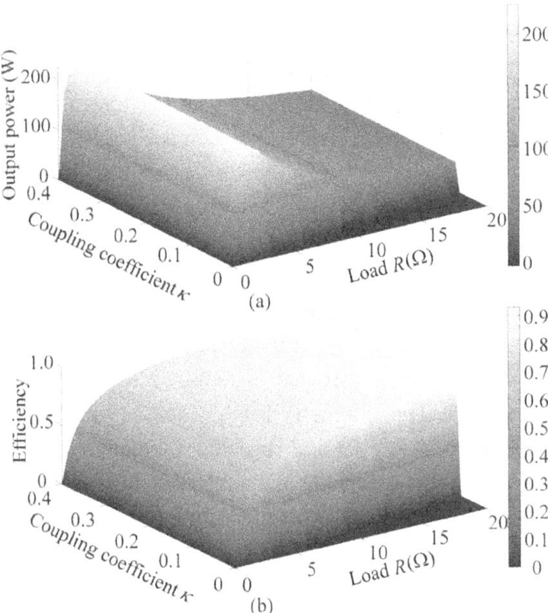

Figure 5.8 System output properties against κ and R: (a) load power; (b) efficiency.

intuitive way the system's output power and efficiency, Fig. 5.8 shows the output characteristics for different coupling coefficients κ and loads R_{eq}. As can be seen, although this PT-WPT system achieves mutual inductance decoupling in the symmetric region, it is still affected by the system load. For practical applications, this battery equivalent load may not be determined in advance and the value of the load varies with the charging process. Therefore, there is a strong need to design a regulation strategy to ensure the robustness of the system under load perturbation.

Based on the challenges that arise in practical applications, a real-time load identification scheme needs to be proposed to achieve robust output for this PT-WPT system under load perturbation. Especially for drone in-flight charging applications, the receiver side should be a lightweight design and avoid the use of wireless communication devices in order to reduce energy losses. This control algorithm to overcome mutual inductance and load perturbation needs to be flexible; namely, the output power should have the ability to be flexibly regulated to adapt to varied charging applications.

In summary, unlike a number of other WPT applications (e.g. electric vehicles [18, 19], wireless motors [20, 21], consumer electronics [22, 23], and wireless lighting [24, 25]), UAV hover charging systems face multiple special technical challenges: how to resist multiple perturbations of mutual inductance and load to achieve output power regulation under different application demands, and how to meet the demand for a lightweight design on the UAV side.

5.3 Output Power Control Based on the Load Identification

5.3.1 Load Identification Method

The above theoretical analysis reveals that the load power and transmission efficiency of the PT-WPT system are not affected by mutual inductance, but are related to the load. Based on Eqs. (5.1)–(5.4), the relationship between load and other system parameters is given as:

$$R_L = \left(\frac{U_p}{I_p} - R_p \right) \frac{L_s}{L_p} - R_s.$$ (5.28)

According to Eq. (5.28), if the parameters of the coupling coils have been determined, system load R_L can be found by detecting the primary current at a given input voltage. This equation shows that the load can be found dynamically without the need to give the initial value of the load, which has wide applicability. After detecting the primary current, the output current can be obtained using Eq. (5.8). Hence, the system load power can be estimated. This scheme does not need to measure the information of the secondary side, which helps to achieve flexible adjustment of the load power on the basis of the transmitting-side control strategy.

To cope with the special challenge of UAV hover charging, a new constant power control scheme based on the PT-symmetric energy mirroring mechanism of the WPT system with dynamic load recognition and no communication link is proposed. The system block diagram is illustrated in Fig. 5.9. Under these double disturbances of coupling strength and load, in order to remove the impact of M on P_{out}, the negative resistance of the PT-WPT system is designed and implemented. In addition, dynamic load recognition is achieved based on the relationship between circuit state variables, and provides a technical solution to the power fluctuation suppression control algorithm.

For this implementation of negative resistance, the basic design principle is that the phase deviation between the equivalent voltage and current across the negative resistance should be maintained at 180 degrees. In this section, a phase synchronization

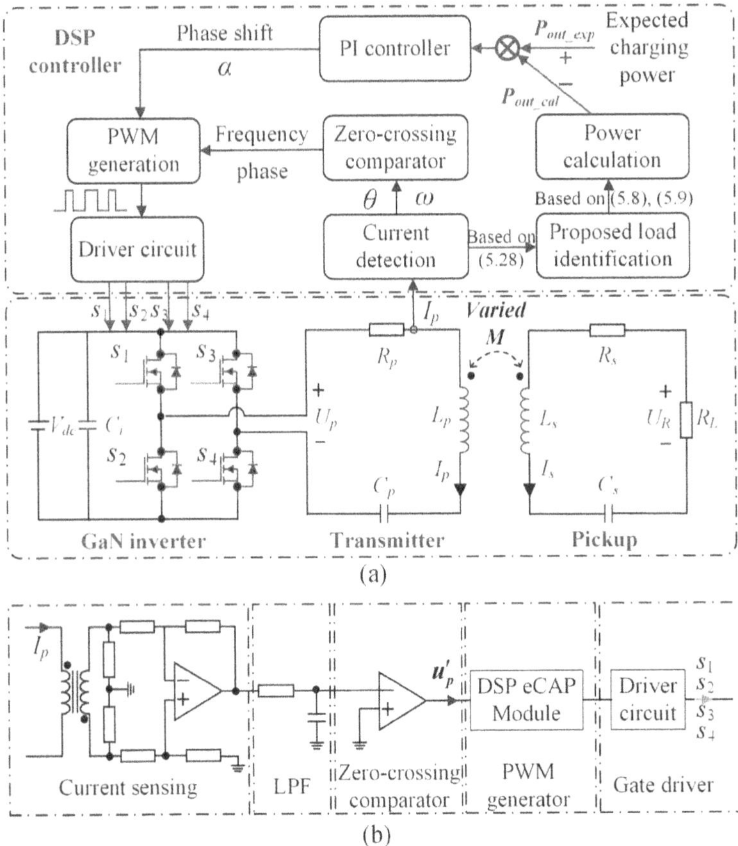

(a)

(b)

Figure 5.9 (a) Control block diagram of the proposed load recognition; (b) hardware implementation circuits.

method (PSM) on the basis of a DSP controller is presented to achieve a negative resistor. The DSP controller generates the drive signals S_1 (or S_2) and S_3 (or S_4) through two kinds of PWM blocks (PWM1 and PWM2). As can be seen in Fig. 5.9(b), this current sensor detects primary-side current I_p, which is passed through an over-zero comparator to produce a wave u'_p with the same phase as I_p. Then, the eCAP module of the DSP will identify the rising edge of u'_p, which contains phase and period information of I_p. Here, a low-pass filter helps to improve system stability and reduce the effects of spurious signals. The desired PWM signal is then produced to control the inverter switching tubes. Based on the above control strategy, the system working angular frequency is auto-stabilized to $\omega_{1,2}$ at this exact PT symmetry range (or ω_0 at this broken PT symmetry range), thus avoiding the influence on system load power of continuous changes in the mutual inductance.

The driving pulse generation process of the proposed PSM is further explained in Fig. 5.10. If this rising edge of u'_p comes, the synchronization pulse SYNCI1 will be

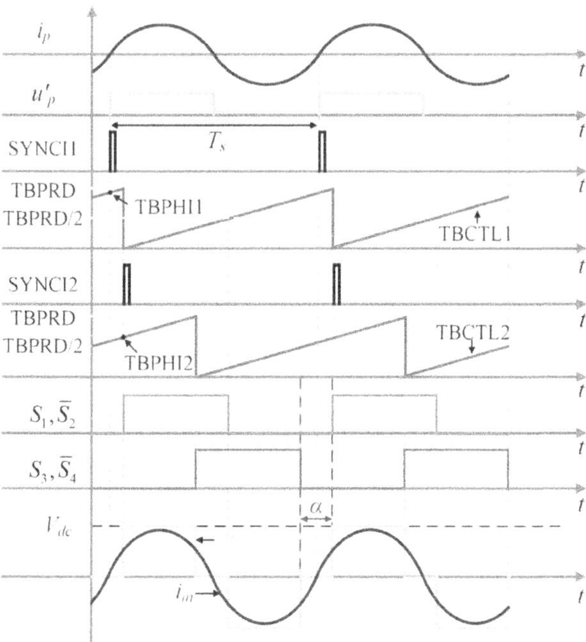

Figure 5.10 The drive signal generation process based on the phase synchronization method.

produced and the counter of PWM1 (i.e. TBCTL1) is immediately updated to
TBPHI1. Here, the counters of PWM1 and PWM2 (TBCTL1 and TBCTL2) are
operating in the up-count state. If TBCTL1 is the same as the maximum value
(TBPRD), S_1 will force the output high (S_2 will force the output low). When
counter TBCTL1 is the same as half of this period (TBPRD/2), S_1 will force the
output low (S_2 will force the output low). Similarly to how S_1 and S_2 are
generated, drive signals S_3 and S_4 can also be produced on the basis of
TBCTL2. However, the distinction between TBCTL1 and TBCTL2 lies in the
different initial values of counters (TBPHI1 and TBPHI2), as shown in Fig. 5.10.
When TBCTL1 reaches its maximum value (TBPRD), the synchronization signal
SYNCI2 is generated and TBCTL2 is immediately updated to the value TBPHI2.
The values of TBPHI1 and TBPHI2 are then set to satisfy Eq. (5.29).
Accordingly, the scheme enables the regulation of the input voltage of the system
to control the load power of the system.

$$
\begin{aligned}
\text{TBPHI1} &= \frac{\alpha \cdot \text{TBPRD}}{4\pi}, \\
\text{TBPHI2} &= \frac{(\pi - \alpha) \cdot \text{TBPRD}}{2\pi},
\end{aligned}
\tag{5.29}
$$

where $\text{TBPRD} = 2\pi/(\omega \text{T}_{\text{CLK}})$ and T_{CLK} represents the duration taken to increase the
counter by 1.

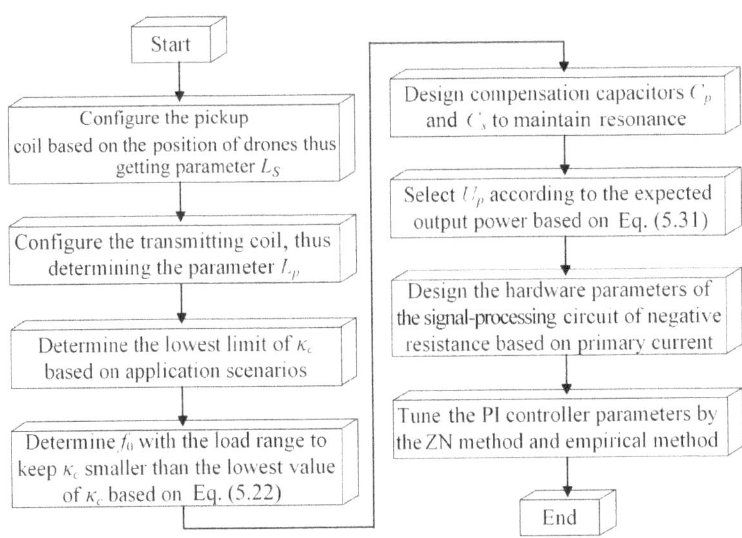

Figure 5.11 Flowchart of UAV system parameter design.

This negative resistance design scheme ensures constant output power under mutual inductance variation, and the impact of load disturbance on system load power can be overcome using this proposed load identification method. At this exact PT symmetry range, this system equivalent load R_L can be determined according to Eq. (5.28). Then, this transmitting-side current can be measured and the load can be calculated from the system parameters. The output current can be computed from this current relation (i.e. Eq. (5.8)) at this exact PT symmetry range. Based on the load current and the recognized load R_L, system load power can be directly calculated. The proposed scheme can achieve the flexible adjustment of the load power of the PT-WPT system and has the benefit of simplicity and reliability without adding any compensation network or power converter.

The parameter design process of this PT-WPT system is shown in Fig. 5.11 for UAV hover charging. It includes the system parameters, the parameters of the negative resistance control circuit, and the parameters of the PI controller. The specific parameter setting process of each step is as follows:

1. The receiving coil can be configured to obtain the parameter value of L_s according to the space available within the drone.
2. The transmitting coil can also be configured to obtain the parameter value of L_p.
3. According to the application scenario, the lowest coupling coefficient of the transmitting and pickup coils measured by the test serves as the desired value of κ_{c_exp}, which ensures the fulfilment of the accurate PT symmetry range for this PT-WPT system. Thus, robust output is guaranteed in the case of changing mutual inductance.

4. According to Eq. (5.22), when a range of load values is obtained, frequency f_0 can be selected and determined so that the value of κ_c becomes smaller than the expected value κ_{c_exp}. Thus, system load power will be kept constant in spite of variations in the mutual inductance, which results in the decoupling of the mutual inductance and the elimination of the effect of M on P_{out}.
5. The compensation capacitors C_p and C_s are calculated and designed according to the resonance equation to maintain the resonant state of the system.
6. The hardware parameters (resistors, capacitors, etc.) of the signal-processing circuits should be designed based on the primary-side current, as shown in Fig. 5.9 (b). For example, the parameters of the operational amplifier (LM6172) differential amplification circuit and the low-pass filter (LPF) circuit should be selected. According to the signal-processing circuit, this negative resistor can be obtained by adaptive frequency adjustment.
7. Based on the determined load, this system power control scheme for load variation requires knowledge of the parameters of the PI controller to obtain the desired dynamic response. Here, the PI controller parameters can be first determined by the Ziegler–Nichols (ZN) scheme and afterwards are fine-tuned through an empirical approach.

Figure 5.12 illustrates the flowchart of the communication-free power fluctuation suppression control algorithm. First, set the inverter work frequency to the resonant frequency ω_0 of this WPT system and then start the system. The PSM-based primary-side current detection and control circuit are utilized to update the system drive signal. In order to ensure that the system obtains stable output characteristics under fluctuations in the UAV position, it is ensured that the value of κ_c is less than the whole coupling coefficient in practical scenarios. In this way, constant output power and efficiency are achieved under mutual inductance variation. The primary-side current is detected and the dynamic load identification is achieved based on Eq. (5.28). The current output power can be obtained based on Eqs. (5.8) and (5.31). The PSM method in the PT-WPT system has achieved constant output power under mutual

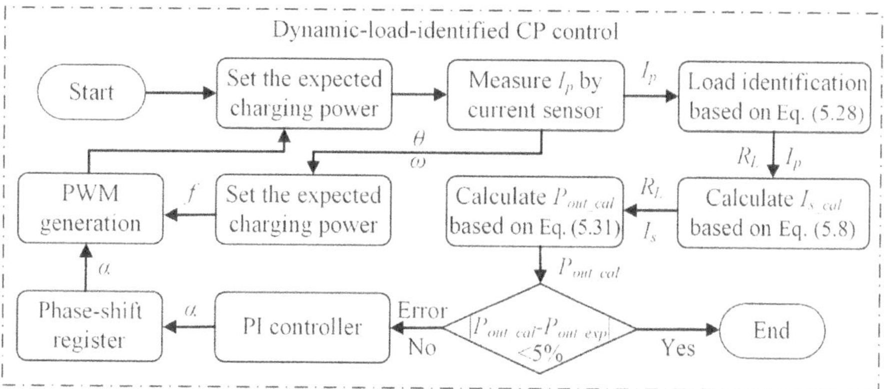

Figure 5.12 The proposed power fluctuation suppression control block diagram.

inductance variation, and the PI controller can overcome the impact of load value changes on the system output power. That is, when the load changes, it results in an adjustment of the phase-shift angle. In conclusion, the proposed system power control scheme on the basis of load recognition is combined with the PSM method to realize a power-regulable PT-WPT system, thus overcoming the double disturbance of mutual inductance and load.

According to the first harmonic approximation (FHA) method, the system input voltage can be represented as Eq. (5.30), and the system load power as Eq. (5.31):

$$U_p = \frac{4}{\sqrt{2}\pi} V_{dc} \cos \frac{\alpha}{2},$$ (5.30)

$$P_{out\_cal} = I_{s\_cal}^2 R_L = \frac{4}{\sqrt{2}\pi} \cos \frac{\alpha}{2} V_{dc} I_p - I_p^2 R_p - \frac{L_p}{L_s} I_p^2 R_s.$$ (5.31)

In order to avoid the effect of resonance deviation in the system, parameter deviation of a single capacitor can be reduced using several capacitors connected in parallel. For example, by connecting eight 1 nF capacitors and eight 100 pF capacitors in parallel, a capacitor with a capacitance value of 8.8 nF can be formed.

In summary, the proposed power fluctuation suppression control only requires measurement and control on the primary side, without wireless communication. Thus, the scheme avoids extra weight on the UAV and so reduces power loss during hovering.

5.3.2   Experimental Verification

The test system (Fig. 5.13) was built to verify this effectiveness of the above control algorithm; the system parameters can be found in Table 5.2. This WPT

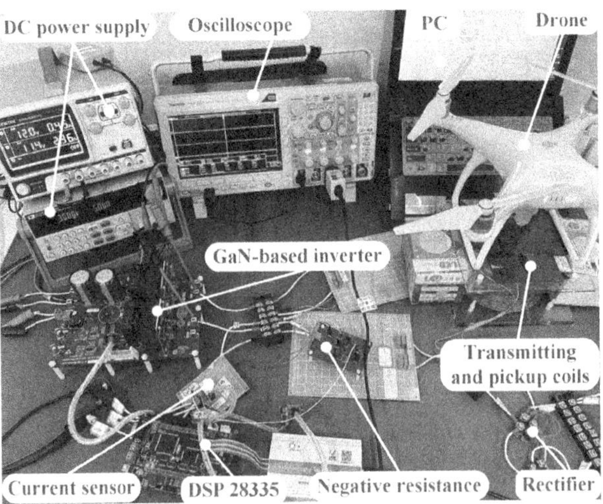

Figure 5.13 Experimental prototype.

Table 5.2 System parameters

Item	Value
Transmitter coil inductance (L_p)	288.45 µH
Transmitter coil resistor (R_p)	0.386 Ω
Transmitter-side capacitor (C_p)	8.782 nF
Receiving coil inductance (L_s)	520.22 µH
Receiving-side resistor (R_s)	0.762 Ω
Receiving-side capacitor (C_s)	4.869 nF
Resistance of load (R_L)	5–20 Ω
Natural resonance frequency (f_0)	100 kHz
DC source voltage (V_{dc})	30 V

system is composed of the energy supply (Itech IT6722A), a GaN inverter (GSP65R13HB-EVB), a signal-processing circuit, a WPT circuit topology, a DSP controller, a rectifier bridge, and a load. The system coils are tightly wound with Litz wire (Φ 0.1 mm $\times$ 200 strands). The compensation capacitors are made of polypropylene capacitors connected in series and parallel. The RF resistors are selected for the load. Fast recovery epitaxial diodes (DSEI2x30–12B) are used to convert the system AC to DC. The current sensor and operational amplifier are CU8965 and LM6172, respectively. The over-zero comparator is a TL3016, which is an ultra-fast low-power comparator with a delay time of only 7.6 ns.

In this proposed system, the load value is determined to be 5–20 Ω based on the UAV battery; the pickup coil is determined to be 520.22 µH; and this system natural resonance frequency is selected to be 100 kHz. Therefore, the maximum value of κ_c must be 0.063 to satisfy the common range of UAV hover charging. This robust output power under mutual inductance variation can be obtained within the selected parameters.

The theoretical, simulation, and experiment results of the system output properties (output power, transmission efficiency, and operating frequency) are illustrated in Figs. 5.14–5.16. When the load value is 15 Ω, the value of κ_c is 0.0482. It can be seen in Fig. 5.14 that if the coupling strength is greater than 0.0482, the system enters the region of accurate PT symmetry, and load power and efficiency are stabilized at 73 W and 91.2%, regardless of the coupling strength disturbance. It can be seen from Fig. 5.15 that the system working frequency deviates from the value of ω_0 and stabilizes at $\omega_{1,2}$ and varies with the coupling coefficient. The experimental simulation and theoretical results are close to each other, which verifies the output properties of this PT-WPT system. Finally, Fig. 5.16 depicts the system load power and efficiency against various loads. It can be observed that the change of load still affects the system output characteristics, so the proposed control algorithm is utilized to achieve flexible regulation of power. The precision of the proposed load detection method and the effectiveness of the power control will now be verified experimentally.

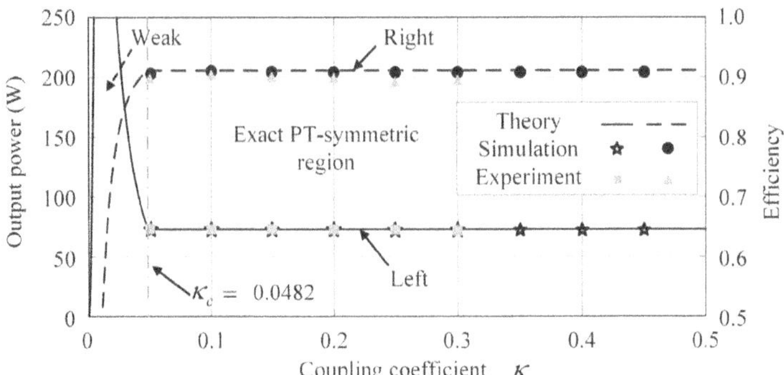

Figure 5.14 System load power and efficiency with different κ.

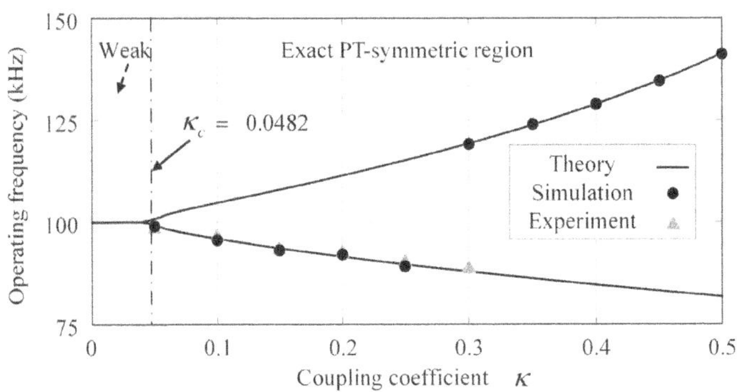

Figure 5.15 System frequency with different κ.

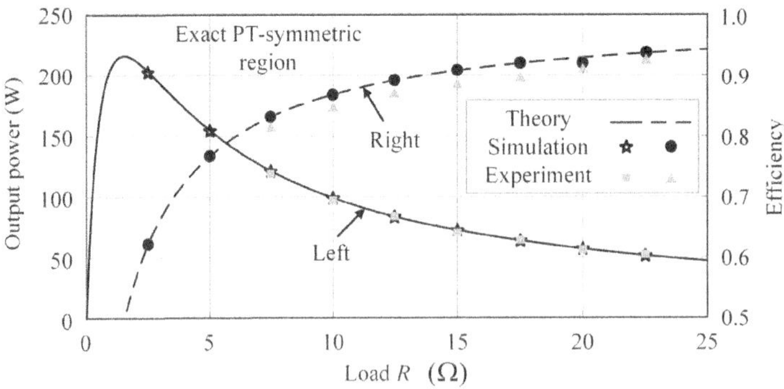

Figure 5.16 System charging power and transfer efficiency with different loads.

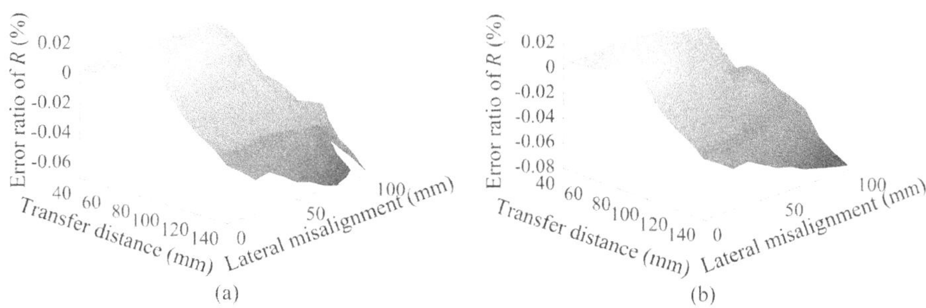

Figure 5.17 Load detection accuracy under different conditions: (a) R_L is 10 Ω; (b) R_L is 15 Ω.

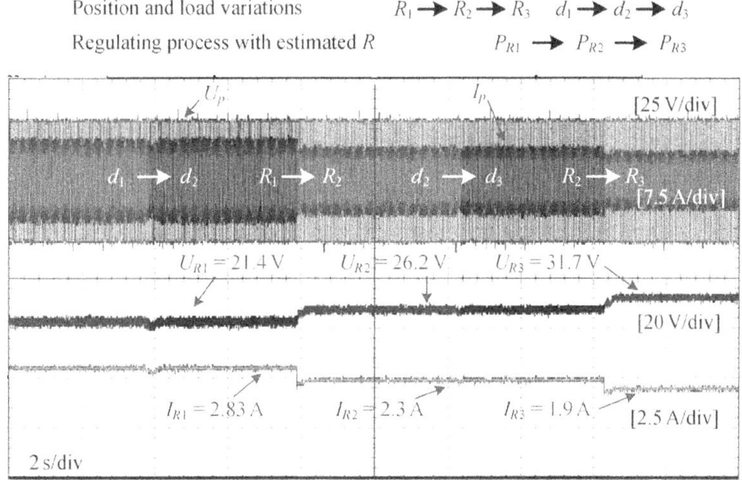

Figure 5.18 Experimental waveforms when M and R_L change.

To validate the precision of the proposed load detection in various environments for drone in-flight charging, the experimental results were tested for different charging distances (40–140 mm), various lateral deviations (0–100 mm), and various battery loads (10 Ω and 15 Ω). The test results for the load detection error rate are plotted in Fig. 5.17. Based on the experimental results, it can be seen that the precision of the load detection is better than 92% – when coupling strength is high, the accuracy can reach more than 96% – which meets the application requirements for UAV hover charging. The load detection scheme only needs to use the transmitting-side current, which is conducive to the design of a communication-free power control scheme on the basis of the primary side only, and meets the need for a lightweight receiver-side design.

To validate the robustness of the power control algorithm against dual perturbation of mutual inductance and load, a dual-perturbation case was designed. The experimental result plots are recorded as shown in Figs. 5.18–5.20. The transmission

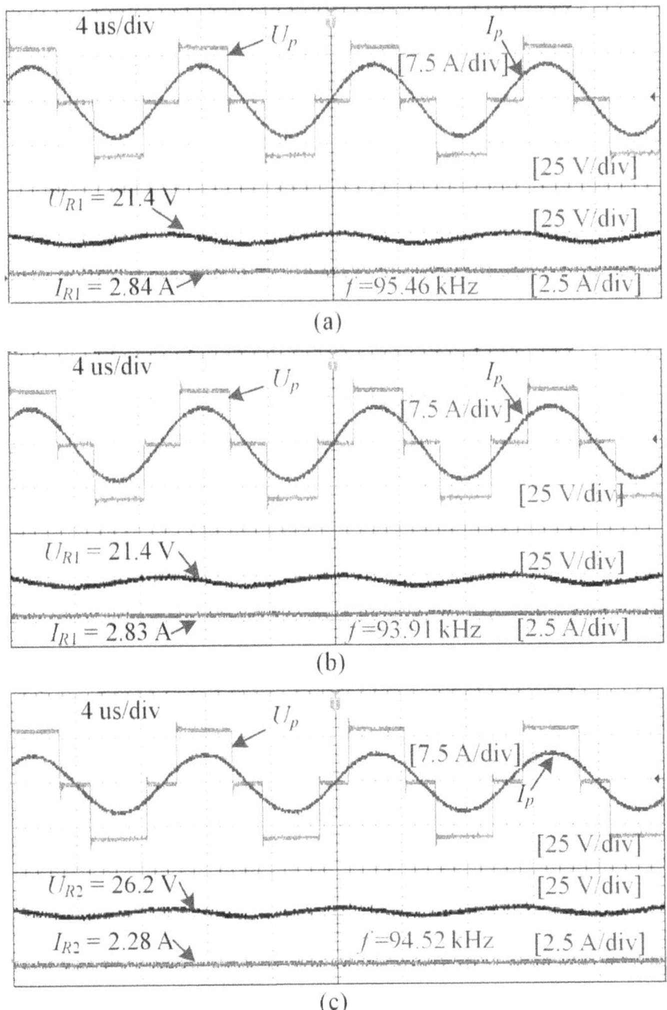

Figure 5.19 Experimental steady-state waveforms: (a) $d_1 = 10$ cm, $R_1 = 7.5\ \Omega$; (b) $d_2 = 9$ cm, $R_1 = 7.5\ \Omega$; (c) $d_2 = 9$ cm, $R_2 = 11.5\ \Omega$.

distance between the coupling coils is changed from 10 cm to 9 cm to 8 cm, and the load varies from 7.5 Ω to 11.5 Ω to 16.5 Ω. The desired load power is 60 W. As can be observed from Fig. 5.18, stable output power can be guaranteed under dual perturbation using the proposed power control scheme. Figures 5.19 and 5.20 depict the steady-state waveform plots under different loads and mutual inductances, demonstrating the effectiveness of this power regulation scheme. The comparison with and without the proposed power control algorithm (Fig. 5.20(b) and (c)) illustrates that the output power is reduced by 19.6% when the control algorithm is not present. These results prove the stability and rapidity of the proposed power fluctuation suppression control algorithm under dual perturbations, which are conducive to fast energy replenishment for the hovering drone.

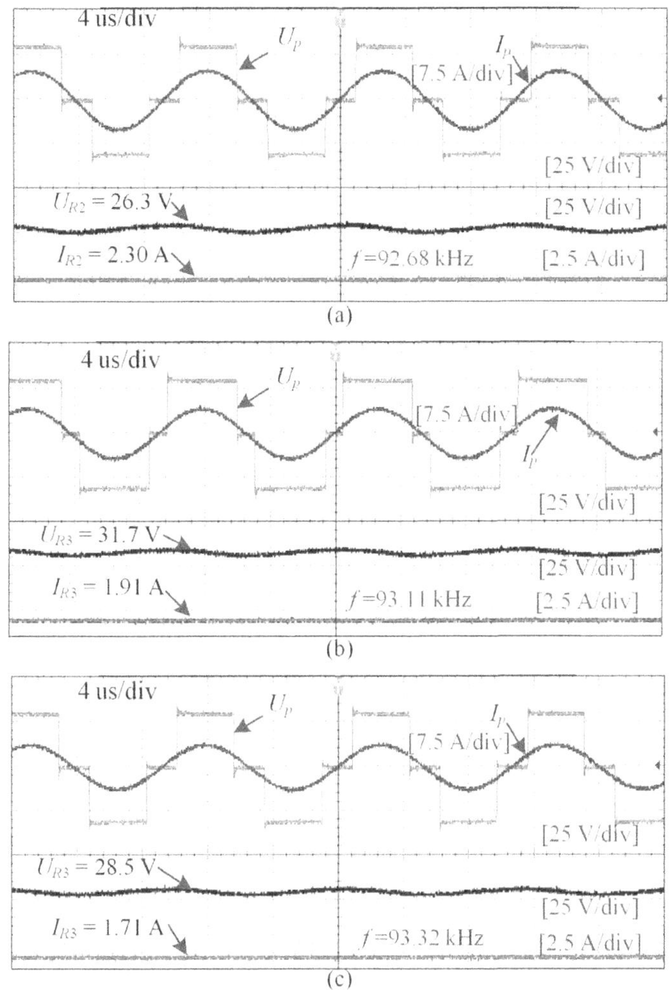

Figure 5.20 Experimental steady-state waveforms. (a) $d_3 = 8$ cm, $R_2 = 11.5$ Ω; (b) $d_3 = 8$ cm, $R_3 = 16.5$ Ω; (c) $d_3 = 8$ cm, $R_3 = 16.5$ Ω, without the proposed control scheme.

References

[1] S. Assawaworrarit, X. Yu, and S. Fan, 'Robust wireless power transfer using a nonlinear parity-time-symmetric circuit', *Nature*, vol. 546, no. 7658, pp. 387–390, Jun. 2017.

[2] J. Zhou, B. Zhang, W. Xiao, D. Qiu, and Y. Chen, 'Nonlinear parity-time-symmetric model for constant efficiency wireless power transfer: application to a drone-in-flight wireless charging platform', *IEEE Trans. Ind. Electron.*, vol. 66, no. 5, pp. 4097–4107, May 2019.

[3] Y. Gu, J. Wang, Z. Liang, and Z. Zhang, 'Mutual-inductance-dynamic-predicted constant current control of LCC-P compensation network for drone wireless in-flight charging', *IEEE Trans. Ind. Electron.*, vol. 69, no. 12, pp. 12710–12719, Dec. 2022.

[4] X. Dai, X. Li, Y. Li, and A. P. Hu, 'Maximum efficiency tracking for wireless power transfer systems with dynamic coupling coefficient estimation', *IEEE Trans. Power Electron.*, vol. 33, no. 6, pp. 5005–5015, Jun. 2018.

[5] Z. Hua, K. T. Chau, W. Han, W. Liu, and T. W. Ching, 'Output-controllable efficiency-optimized wireless power transfer using hybrid modulation', *IEEE Trans. Ind. Electron.*, vol. 69, no. 5, pp. 4627–4636, May 2022.

[6] Z. Zhang, S. Shen, Z. Liang, S. H. K. Eder, and R. Kennel, 'Dynamic-balancing robust current control for wireless drone-in-flight charging', *IEEE Trans. Power Electron.*, vol. 37, no. 3, pp. 3626–3635, Mar. 2022.

[7] K. Chen and Z. Zhang, 'Rotating-coordinate-based mutual inductance estimation for drone in-flight wireless charging systems', *IEEE Trans. Power Electron.*, vol. 38, no. 9, pp. 11685–11693, Sept. 2023.

[8] Y. Gu, J. Wang, Z. Liang, and Z. Zhang, 'A wireless in-flight charging range extended PT-WPT system using S/single-inductor-double-capacitor compensation network for drones', *IEEE Trans. Power Electron.*, vol. 38, no. 10, pp. 11847–11858, Oct. 2023.

[9] M. Kiani and M. Ghovanloo, 'The circuit theory behind coupled-mode magnetic resonance-based wireless power transmission', *IEEE Trans. Circuits Syst. I: Reg. Papers*, vol. 59, no. 9, pp. 2065–2074, Sept. 2012.

[10] Y. Gu, Q. Zhu, L. Jia, G. Li, and Z. Zhang, 'Optimized high-order compensation topology for PT-WPT systems with expanded constant power region', *IEEE Trans. Magn.*, 2023, doi: 10.1109/TMAG.2023.3287481.

[11] C. Rong, B. Zhang, Z. Wei, L. Wu, and X. Shu, 'A wireless power transfer system for spinal cord stimulation based on generalized parity–time symmetry condition', *IEEE Trans. Ind. Appl.*, vol. 58, no. 1, pp. 1330–1339, Jan.–Feb. 2022.

[12] Y. Gu, J. Chen, S. Chang, and Z. Zhang, 'Constant power control against M/R with expanded PT-symmetric range for wireless in-flight charging of drones', *IEEE Trans. Magn.*, 2023, doi: 10.1109/TMAG.2023.3284826.

[13] Y. Gu, J. Wang, Z. Liang, and Z. Zhang, 'Communication-free power control algorithm for drone wireless in-flight charging under dual-disturbance of mutual inductance and load', *IEEE Trans. Ind. Inform.*, vol. 20, no. 3, pp. 3703–3714, Mar. 2024.

[14] R. Huang and B. Zhang, 'Frequency, impedance characteristics and HF converters of two-coil and four-coil wireless power transfer', *IEEE J. Emerg. Sel. Topics Power Electron.*, vol. 3, no. 1, pp. 177–183, Mar. 2015.

[15] S. Cheon, Y.-H. Kim, S.-Y. Kang, M. L. Lee, J.-M. Lee, and T. Zyung, 'Circuit-model-based analysis of a wireless energy-transfer system via coupled magnetic resonances', *IEEE Trans. Ind. Electron.*, vol. 58, no. 7, pp. 2906–2914, Jul. 2011.

[16] H. A. Haus, *Waves and Fields in Optoelectronics*. Prentice-Hall, 1984.

[17] J. A. Sanders, F. Verhulst, and J. A. Murdock, *Averaging Methods in Nonlinear Dynamical Systems*, 2nd ed. Springer, 2007.

[18] P. Zhang, M. Saeedifard, O. C. Onar, Q. Yang, and C. Cai, 'A field enhancement integration design featuring misalignment tolerance for wireless EV charging using LCL topology', *IEEE Trans. Power Electron.*, vol. 36, no. 4, pp. 3852–3867, Apr. 2021.

[19] X. Dai, J. Jiang, and J. Wu, 'Charging area determining and power enhancement method for multiexcitation unit configuration of wirelessly dynamic charging EV system', *IEEE Trans. Ind. Electron.*, vol. 66, no. 5, pp. 4086–4096, May 2019.

[20] C. Jiang, K. T. Chau, C. H. T. Lee, W. Han, W. Liu, and W. H. Lam, 'A wireless servo motor drive with bidirectional motion capability', *IEEE Trans. Power Electron.*, vol. 34, no. 12, pp. 12001–12010, Dec. 2019.

[21] W. Han, K. T. Chau, Z. Hua, and H. Pang, 'An integrated wireless motor system using laminated magnetic coupler and commutative-resonant control', *IEEE Trans. Ind. Electron.*, vol. 69, no. 5, pp. 4342–4352, May 2022.

[22] Y. Gu, J. Wang, Z. Liang, Y. Wu, C. Cecati, and Z. Zhang, 'Single-transmitter multiple-pickup wireless power transfer: advantages, challenges, and corresponding technical solutions', *IEEE Ind. Electron. Mag.*, vol. 14, no. 4, pp. 123–135, Dec. 2020.

[23] C. Liu, C. Jiang, J. Song, and K. T. Chau, 'An effective sandwiched wireless power transfer system for charging implantable cardiac pacemaker', *IEEE Trans. Ind. Electron.*, vol. 66, no. 5, pp. 4108–4117, May 2019.

[24] X. Qu, W. Zhang, S.-C Wong, and C. K. Tse, 'Design of a current-source-output inductive power transfer LED lighting system', *IEEE J. Emerg. Sel. Topics Power Electron.*, vol. 3, no. 1, pp. 306–314, Mar. 2015.

[25] W. Liu, K. T. Chau, C. H. T. Lee, C. Jiang, W. Han, and W. H. Lam, 'A wireless dimmable lighting system using variable-power variable-frequency control', *IEEE Trans. Ind. Electron.*, vol. 67, no. 10, pp. 8392–8404, Oct. 2020.

6   Charging Range Extension for the PT-Symmetric WPT System

6.1   Introduction

Because of the limitation of battery capacity, it is difficult for UAVs to achieve wide-ranging and continuous operation in complex environments. In recent years, wireless power transfer (WPT) has been able to provide contact-free powered replenishment for UAVs, with the advantages of flexibility, safety, and efficiency [1, 2], thus overcoming the problem of limited UAV range. Considering some special application scenarios (lack of landing plates), in-flight charging [3–6] can allow drone charging while the drone remains operational. A great challenge for this system is achievement of stable charging for the UAV battery in the case of perturbations of the UAV's attitude and position.

For the common magnetically coupled resonant (MCR) WPT system, there are a few options used to resist changes to the load power caused by changes in the relative positions of the system's coils. These schemes can mainly be divided into: coil structure design [7, 8], parameter prediction [9, 10], impedance matching network [11, 12], and parity–time (PT) symmetry theory [13, 14]. As discussed already, the PT-WPT system is an emerging technology that can achieve adaptive constant power (CP) output with the help of a control structure on the transmitter side. This system is suited for applications requiring lightweight design on the receiver side (e.g. UAV hover charging).

Based on the modelling and analysis in Chapter 5, load power P_{out} and efficiency η of the PT-WPT system have been compared with those of the common MCR-WPT system, illustrated in Fig. 6.1. As can be clearly observed, if this coupling strength is greater than the threshold value, the PT-WPT system enters the exact PT region with higher output power than the MCR-WPT system without significant loss of transmission efficiency. In this region, the PT-WPT system achieves constant load power and efficiency under the variation of κ. Since the control circuits are all mounted on the transmitting side, the system reduces the weight of the receiving side and so avoids extra power consumption while the UAV hovers. The requirement for this PT-WPT system to satisfy the CP output is that κ is greater than the critical coupling κ_c. Therefore, there is a need for a scheme to achieve robust charging power range expansion that ensures a large range of stable charging powers in various complex scenarios.

To extend the transmission range for wireless charging, common methods mainly include: repeater coils [15, 16], magnetic materials [17], and parameter optimization

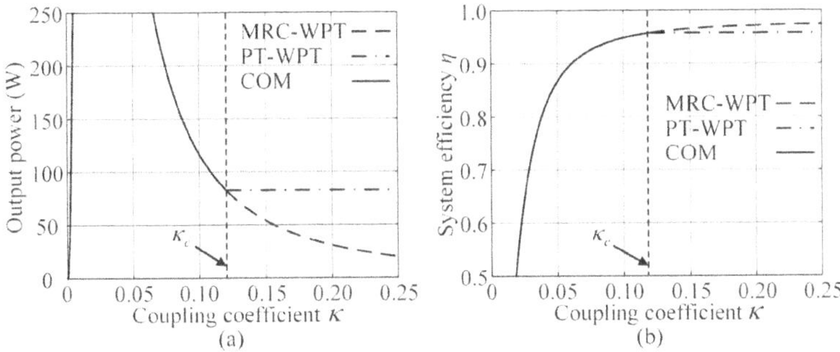

Figure 6.1 (a) System load power; (b) system transfer efficiency.

[18, 19]. A PT-WPT system with multiple relay coils has been proposed and demonstrated to have lower critical coupling coefficients and larger CP range [16]. However, the system requires the same coupling coefficients from the relay coils to the transmitter and receiver coils, which presents a greater challenge for practical applications. Magnetic materials, including ferrite and nanocrystals, can achieve magnetic field convergence and thus enhance magnetic coupling intensity of the system coupling coils [17]. However, when expanding this charging range the resulting power losses and additional weight should be considered, which is especially relevant to the UAV case. For the PT-WPT system, the addition of this receiving-side inductor would reduce κ_c [18], thus expanding the CP range of the system. In practice, the space and load on the pickup side are limited, and the size of the pickup coil cannot be increased infinitely to expand the CP region. One attempt at a PT-WPT system with additional inductors on the receiving side achieved a 30.4% expansion of the CP transfer range and enhanced the system's degree of freedom [19]. However, the CP region is not adjustable with these fixed system parameters. Therefore, there is a strong need to flexibly expand and adaptively adjust the robust charging region of PT-WPT systems, and at the same time minimize the extra weight on the receiving side to accommodate different models of UAV for wireless charging.

In summary, some advances have been made in the research on PT-WPT systems. The design of the primary-side feedback circuit can effectively negate the problem of system output power fluctuation under variation in the location of the receiving coil, which provides a new solution to this CP output requirement for UAV hover charging systems. However, the challenges faced by drone in-flight charging systems are more complex than other common wireless charging applications, especially the challenges of persistent perturbations in the UAV attitude, maintaining a UAV-side lightweight design, and providing flexible regulation of charging power. There is a need to explore and solve the problem of expanding and adaptive regulation of PT-symmetric regions in order to cope with these challenges.

This chapter focuses on the CP region extension and regulation methods for PT-WPT systems. First, a transmitting series (S) and receiving single inductor double capacitor (SLDC) compensation network is introduced, which can greatly extend and regulate the CP area of PT-WPT systems and provide improved misalignment tolerance. Then, a novel flexible CP range extension method is introduced in detail, which has the major advantages of long-range robustness, high flexibility, and a lightweight receiving side.

6.2 Charging Range Extension with S/SLDC High-Order Topology

Compared to other technical solutions [20–25], this PT-WPT system could provide robust charging power by simple primary-side control only. However, according to the theoretical analysis of this PT-WPT system in the previous section, robust output power can only be achieved if κ is larger than the specific range (namely the critical coupling coefficient κ_c) [26]. On the contrary, if the system is weakly coupled, the system load power and efficiency are difficult to stabilize automatically when there is variation in the locations of the coupling coils. Therefore, the PT symmetry (robust output) range needs to be extended to achieve longer transfer distances to satisfy the demand of practical working environments. This section focuses on the S/SLDC compensation network [27] – which can achieve great expansion of the CP region – in the context of the working requirements for UAV hover charging.

6.2.1 System Analysis and Modelling

This proposed PT-WPT system of the S/SLDC compensation network is shown in Fig. 6.2. V_{dc} is the input DC voltage that is passed through the full-bridge inverter constituted by S_1–S_4 to generate the AC power required by the WPT system. The primary-side series topology consists of primary-side inductor L_p and series capacitor C_p, and the secondary-side S/SLDC topology consists of secondary-side inductor L_s, increased inductance L_r, and dual capacitors C_{s1} and C_{s2}. The primary-side internal resistance is R_p, the secondary-side internal resistance is R_s, and mutual inductance of the coupling coils can be expressed as M_{ps}. The UAV battery load R_{dro} is rectified and filtered to be equivalent to the load R_L, where $R_L = 8R_{dro}/\pi^2$. The key to realizing this PT-WPT system through this control circuit design is to ensure that this phase of the inverter output voltage u_p is opposite to the primary-side current $-i_p$.

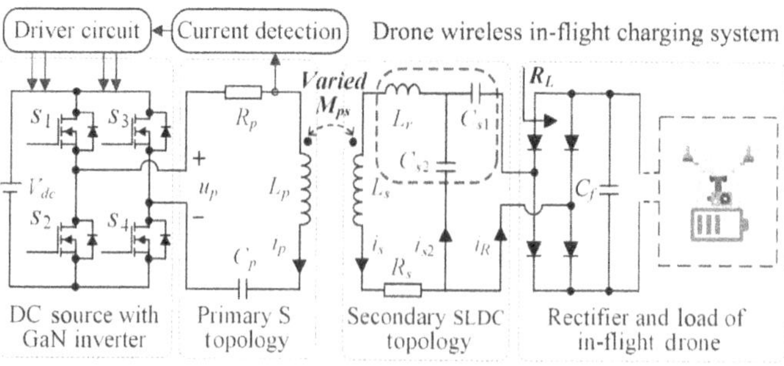

Figure 6.2 System schematic of the proposed S/SLDC compensation network.

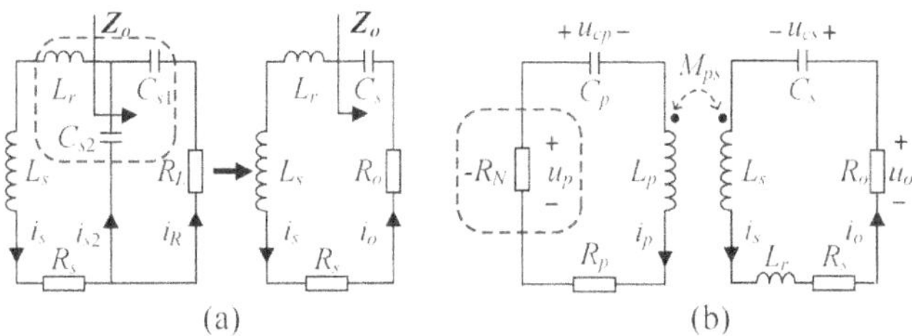

Figure 6.3 (a) Receiving-side equivalent conversion; (b) system simplified circuit.

The coupled-mode theory (CMT) model has been used to analyse the proposed S/SLDC compensation network, which has been simplified to make it easier to analyse the energy flow. According to the equivalent circuit method, the double capacitances C_{s1} and C_{s2} on the secondary side can be simplified to a single capacitance C_s, as illustrated in Fig. 6.3(a). The equivalent input impedance Z_o is given by:

$$Z_o = \frac{1}{j\omega C_s} + R_o = \frac{1}{j\omega C_{s2}} \,//\, \left(\frac{1}{j\omega C_{s1}} + R_L\right)$$
$$= \frac{\left[C_{s1} + C_{s2} + \omega^2 C_{s1s}^2 C_{s2} R_L^2\right]/(j\omega) + C_{s1}^2 R_L}{(C_{s1} + C_{s2})^2 + \omega^2 C_{s1}^2 C_{s2}^2 R_L^2}. \tag{6.1}$$

In this WPT system, it could be fulfilled that Eq. (6.2) is

$$\omega^2 C_{s1}^2 C_{s2}^2 R_L^2 \ll (C_{s1} + C_{s2})^2 \tag{6.2}$$

and Eq. (6.3) is

$$\omega^2 C_{s1}^2 C_{s2} R_L^2 \ll C_{s1} + C_{s2}. \tag{6.3}$$

Simplifying Eq. (6.1), it can be written as:

$$C_s = C_{s1} + C_{s2},$$
$$R_o = \frac{C_{s1}^2 R_L}{(C_{s1} + C_{s2})^2} - \lambda^2 R_L. \tag{6.4}$$

In Eq. (6.4), $\lambda = C_{s1}/(C_{s1} + C_{s2})$ is the capacitance ratio, which ranges from 0 to 1. The equivalent compensation capacitance and the equivalent load are denoted as C_s and R_o, respectively.

In order to prove the reasonableness of the above assumptions, the following experimental parameters used:

- $L_p = 73.5\ \mu H$
- $L_s = 140\ \mu H$
- $C_{s1} = 7.55\ nF$
- $C_{s2} = 5.04\ nF$
- $f_0 = 100\ kHz$
- $R_L = 15\ \Omega.$

If κ is 0.5, the largest value of this calculated operating frequency ω is ~$2\pi \times 125$ kHz. The left-hand side of Eq. (6.2) has a value of only $2.01\mathrm{E}^{-19}$, while the right-hand side has a value of $1.59\mathrm{E}^{-16}$, which is about 800 times that of the former. Therefore, it is reasonable to assume that Eq. (6.2) is within a certain error range. The left-hand side value of Eq. (6.3) is only $3.99\mathrm{E}^{-11}$, and the right-hand side is $1.26\mathrm{E}^{-08}$, which is about 320 times the former. Therefore, based on the numerical comparison, it is also reasonable to assume that Eq. (6.3) is within a certain error range.

According to the equivalence transformation illustrated in Fig. 6.3(a), this pickup-side SLDC compensation network is equivalent to the LC network. Therefore, the proposed PT-WPT system with S/SLDC topology can be illustrated as in Fig. 6.3(b). The system circuit equation is:

$$
\begin{aligned}
&L_p \frac{di_p}{dt} + M_{ps} \frac{di_s}{dt} + u_{cp} + i_p R_p = u_p, \\
&M_{ps} \frac{di_p}{dt} + (L_s + L_r) \frac{di_s}{dt} + u_{cs} + i_s(R_s + R_o) = 0, \\
&C_p \frac{du_{cp}}{dt} = i_p, \\
&C_s \frac{du_{cs}}{dt} = i_s.
\end{aligned}
\tag{6.5}
$$

In the above equations, i_p, i_s, and i_o represent these currents across L_p, L_s, and R_o, respectively. u_{cp} and u_{cs} stand for voltages of the coupling coils L_p and L_s, respectively. Here, this system input voltage is $u_p = 4V_{dc}/\sqrt{2}\pi$. Based on this CMT model [28], energy ψ_p and ψ_s of the primary and secondary resonators are defined as:

$$
\begin{aligned}
\psi_p &= \Psi_p e^{j(\omega t + \theta_p)} = \sqrt{\frac{L_p}{2}} i_p + j\sqrt{\frac{C_p}{2}} u_{cp}, \\
\psi_s &= \Psi_s e^{j(\omega t + \theta_s)} = \sqrt{\frac{L_s + L_r}{2}} i_s + j\sqrt{\frac{C_s}{2}} u_{cs}.
\end{aligned}
\tag{6.6}
$$

Here, ω denotes the system working angular frequency, Ψ_p and Ψ_s denote the energy amplitudes, and θ_p and θ_s denote the relative phases. In this way, the system currents and voltages are represented as:

$$
\begin{aligned}
i_p &= \Psi_p \sqrt{\frac{2}{L_p}} \cos(\omega t + \theta_p), \\
u_{cp} &= \Psi_p \sqrt{\frac{2}{C_p}} \sin(\omega t + \theta_p), \\
i_s &= \Psi_s \sqrt{\frac{2}{L_s + L_r}} \cos(\omega t + \theta_s), \\
u_{cs} &= \Psi_s \sqrt{\frac{2}{C_s}} \sin(\omega t + \theta_s).
\end{aligned}
\tag{6.7}
$$

The primary and secondary resonance frequencies can be represented as $\omega_p = 1/\sqrt{L_p C_p}$ and $\omega_s = 1/\sqrt{(L_s + L_r)C_s}$, respectively. For the PT-WPT system, it should be satisfied that $\omega_p = \omega_s = \omega_0$. Then, substituting Eq. (6.7) into Eq. (6.5), the system dynamic equations based on the CMT model are concluded as shown in Eq. (6.8) using the averaging method [29]. According to Eq. (6.6), the derivatives of energy modes are expressed as Eq. (6.9). After that, replacing Eq. (6.8) into Eq. (6.9) and neglecting high-order items, the system dynamic equations are expressed as in Eq. (6.10):

$$
\begin{aligned}
\frac{d\Psi_p}{dt} &= \frac{1}{1-k'^2}\left[\frac{2V_{dc}}{\pi\sqrt{2L_p}} - \frac{R_p}{2L_p}\Psi_p + \frac{k'(R_s + R_o)}{2(L_s + L_r)}\right.\\
&\quad \left. \times \Psi_s\cos(\theta_p - \theta_s) - \frac{k'\omega_s}{2}\Psi_s\sin(\theta_p - \theta_s)\right],\\
\Psi_p\left(\omega + \frac{d\theta_p}{dt}\right) &= \frac{\omega_p}{2}\Psi_p + \frac{1}{1-k'^2}\left[\frac{\omega_p}{2}\Psi_p - \frac{k'\omega_s}{2}\right.\\
&\quad \left. \times \Psi_s\cos(\theta_p - \theta_s) - \frac{k'(R_s + R_o)}{2(L_s + L_r)}\Psi_s(\theta_p - \theta_s)\right],\\
\frac{d\Psi_s}{dt} &= \frac{1}{1-k'^2}\left[-\frac{2k'V_{dc}}{\pi\sqrt{2L_p}}\cos(\theta_p - \theta_s) - \frac{R_s + R_o}{2(L_s + L_r)}\Psi_s\right.\\
&\quad \left. + \frac{k'R_p}{2L_p}\Psi_p\cos(\theta_p - \theta_s) + \frac{k'\omega_p}{2}\Psi_p\sin(\theta_p - \theta_s)\right],\\
\Psi_s\left(\omega + \frac{d\theta_s}{dt}\right) &= \frac{\omega_s}{2}\Psi_s + \frac{1}{1-k'^2}\left[-\frac{2k'V_{dc}}{\pi\sqrt{2L_p}}\sin(\theta_p - \theta_s)\right.\\
&\quad \left. + \frac{\omega_s}{2}\Psi_s - \frac{k'\omega_p}{2}\Psi_p\cos(\theta_p - \theta_s) + \frac{k'R_p}{2L_p}\Psi_p\sin(\theta_p - \theta_s)\right].
\end{aligned}
$$

(6.8)

$$
\begin{aligned}
\frac{d\psi_p}{dt} &= \frac{1}{\Psi_p}\frac{d\Psi_p}{dt}\psi_p + j\left(\omega + \frac{d\theta_p}{dt}\right)\psi_p,\\
\frac{d\psi_s}{dt} &= \frac{1}{\Psi_s}\frac{d\Psi_s}{dt}\psi_s + j\left(\omega + \frac{d\theta_s}{dt}\right)\psi_s.
\end{aligned}
$$

(6.9)

$$
\frac{d}{dt}\begin{bmatrix}\psi_p\\\psi_s\end{bmatrix} = \begin{bmatrix} j\omega_0 + \dfrac{1}{\Psi_p}\dfrac{2V_{dc}}{\pi\sqrt{2L_p}} - \dfrac{R_p}{2L_p} & -j\kappa'\dfrac{\omega_0}{2}\\[2ex] -j\kappa'\dfrac{\omega_0}{2} & j\omega_0 - \dfrac{R_s + R_o}{2(L_s + L_r)} \end{bmatrix}\begin{bmatrix}\psi_p\\\psi_s\end{bmatrix}.
$$

(6.10)

Here, the coupling coefficient of the coupling coils has been defined as $\kappa = M_{ps}/\sqrt{L_p L_s}$ and the equivalent coupling coefficient of the proposed system has been defined as $\kappa' = M_{ps}/\sqrt{L_p(L_s + L_r)}$. According to Eq. (6.10), the system frequency property equation is described as:

$$
\begin{bmatrix} j(\omega_0 - \omega) + \dfrac{1}{\Psi_p}\dfrac{2V_{dc}}{\pi\sqrt{2L_p}} - \dfrac{R_p}{2L_p} & -j\kappa'\dfrac{\omega_0}{2}\\[2ex] -j\kappa'\dfrac{\omega_0}{2} & j(\omega_0 - \omega) - \dfrac{R_s + R_o}{2(L_s + L_r)} \end{bmatrix} = 0.
$$

(6.11)

By separating the real and imaginary terms of the above equation, Eq. (6.11) can be rewritten as:

$$\frac{\kappa'^2\omega_0^2}{4} - (\omega - \omega_0)^2 + \left(\frac{R_p}{2L_p} - \frac{1}{\Psi_p}\frac{2V_{dc}}{\pi\sqrt{2L_p}}\right) \times \frac{R_s + R_o}{2(L_s + L_r)} = 0,$$

$$(\omega - \omega_0)\left(\frac{1}{\Psi_p}\frac{2V_{dc}}{\pi\sqrt{2L_p}} - \frac{R_p}{2L_p} - \frac{R_s + R_o}{2(L_s + L_r)}\right) = 0.$$

(6.12)

The steady-state solution of ω can be obtained as Eq. (6.13), and these system energy amplitudes are represented as Eq. (6.14):

$$\omega = \omega_{1,2} = \omega_0 \pm \sqrt{\left(\frac{\kappa'\omega_0}{2}\right)^2 - \left[\frac{R_s + R_o}{2(L_s + L_r)}\right]^2},$$

(6.13)

$$\Psi_p = \Psi_s = \frac{2\sqrt{2L_p}(L_s + L_r)V_{dc}}{\pi\left[R_p(L_s + L_r) + (R_s + R_o)L_p\right]}.$$

(6.14)

Based on Eq. (6.13), the value of κ_c in the proposed system can be obtained as Eq. (6.15). If the coupling coefficient $\kappa \geq \kappa_c$, the system will work at this exact PT symmetry range. On the contrary, the system would operate at the broken PT symmetry range with the same output properties as the conventional WPT system at the frequency ω_0:

$$\kappa_c = \frac{R_s + R_o}{\omega_0\sqrt{L_s(L_s + L_r)}} = \frac{R_s + \lambda^2 R_L}{\omega_0\sqrt{L_s(L_s + L_r)}}.$$

(6.15)

It is worth noting that this value of κ_c depends significantly on the frequency ω_0, capacitance ratio λ, system load R_L, receiving coil L_s, and additional coil L_r. λ and L_r are the extra degrees of freedom of the proposed S/SLDC compensation network, and the value of κ_c could be decreased easily and flexibly with the parameter design. With the reduction of κ_c, the proposed system could achieve the expansion of this exact PT symmetry range to obtain a large area of robust charging features.

At this exact PT symmetry range, according to Eqs. (6.13) and (6.14), the working frequency will be automatically stabilized at $\omega_{1,2}$ instead of ω_0; the latter would vary with the change of κ. At the same time, the system load power P_{out} and efficiency η are expressed as:

$$P_{out} = \frac{R_o\Psi_s^2}{L_s + L_r} = \frac{8V_{dc}^2 R_o L_p(L_s + L_r)}{\pi^2\left[R_p(L_s + L_r) + (R_s + R_o)L_p\right]^2},$$

(6.16)

$$\eta = \frac{R_o\Psi_s^2/(L_s + L_r)}{R_p\Psi_p^2/L_p + (R_s + R_o)\Psi_s^2/(L_s + L_r)} = \frac{R_o}{R_p(L_s + L_r)/L_p + R_s + R_o}.$$

(6.17)

From Eqs. (6.16) and (6.17) it can be seen that system load power and transmission efficiency are not related to κ in the region of exact PT symmetry. Therefore, a constant output current or output voltage will be kept at this exact PT symmetry range

Table 6.1 Comparison of system output features

	proposed system	SS PT-WPT system
Critical coupling coefficient	$\dfrac{R_s + \lambda^2 R_L}{\omega_0 \sqrt{L_s(L_s + L_r)}}$	$\dfrac{R_s + R_L}{\omega_0 L_s}$
Output power	$\dfrac{8V_{dc}^2 \lambda^2 R_L L_p (L_s + L_r)/\pi^2}{\left[R_p(L_s + L_r) + (R_s + \lambda^2 R_L)L_p\right]^2}$	$\dfrac{8V_{dc}^2 R_L L_p L_s/\pi^2}{\left[R_p L_s + (R_s + R_L)L_p\right]^2}$
Transfer efficiency	$\dfrac{\lambda^2 R_L}{R_p(L_s + L_r)/L_p + R_s + \lambda^2 R_L}$	$\dfrac{R_L}{R_p L_s/L_p + R_s + R_L}$

even if mutual inductance changes. Table 6.1 shows the differences in the output
features between the PT-WPT system of the proposed S/SLDC compensation network
and an S/S compensation network. As can be seen, in comparison to the common S/S
compensation network, κ_c of this proposed system will decrease with λ and L_r, which
enlarges this robust system charging range. Therefore, despite the wide and continuous variation of the air gap between the transmitting device and hovering UAV, the
proposed system allows the UAV to obtain an enlarged CP charging region, thus
effectively enhancing the robustness and adaptability of UAV hover charging.

6.2.2 System Output Characteristics

On the basis of previous theoretical analysis, this PT-WPT system with the proposed
S/SLDC compensation network offers two extra degrees of freedom (λ and L_r) in
comparison to the S/S compensation network, and the system parameters can be
flexibly designed to decrease κ_c. Although this reduction of κ_c could extend the
system's CP range, it should be noted that the system efficiency η is influenced by λ
and L_r. Therefore, in order to optimally design this proposed system, the effects of
these critical system parameters on κ_c and η are analysed. The relevant profiles are
shown in Figs. 6.4–6.6, and are useful in guiding system parametric design of the S/
SLDC compensation network.

According to Eqs. (6.15) and (6.17), the system output characteristics, κ_c versus η,
mostly depend on the system load, the capacitance ratio, and the additional inductance
when the transmitting and receiving coils are designed. The curves of system output
characteristics versus load for different capacitance ratios are depicted in Fig. 6.4,
which indicates that a bigger load to capacitance ratio will result in higher transfer
efficiency. The load of 15 Ω was chosen and fixed for the next analysis in order to
further explore the system output properties. From Fig. 6.5 it can be seen that κ_c will
reduce as the capacitance ratio decreases, but system efficiency also reduces simultaneously. When this additional inductance is added, κ_c will decrease along with the
efficiency, as shown in Fig. 6.6. Since this PT-WPT system is expected to have a
lower critical coupling coefficient and higher output efficiency, in order to achieve the
extension of the PT symmetry range, both critical coupling and efficiency need to be
taken into account to design the appropriate capacitance ratio and additional
inductance.

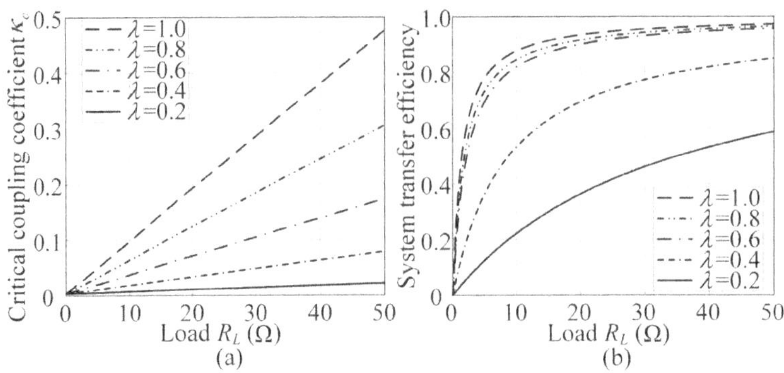

Figure 6.4 (a) κ_c and (b) η under different load R_L and λ.

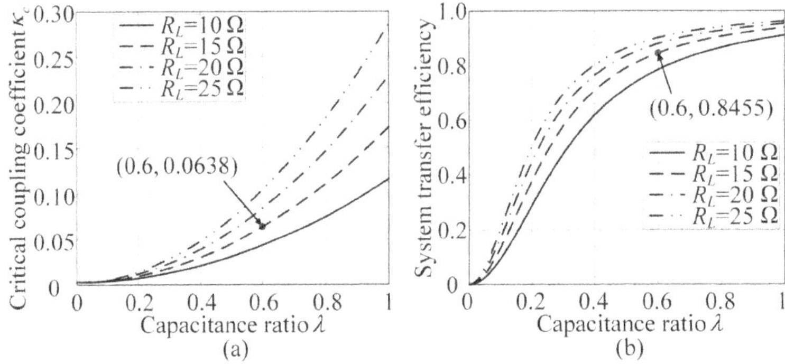

Figure 6.5 (a) κ_c and (b) η under different λ and R_L.

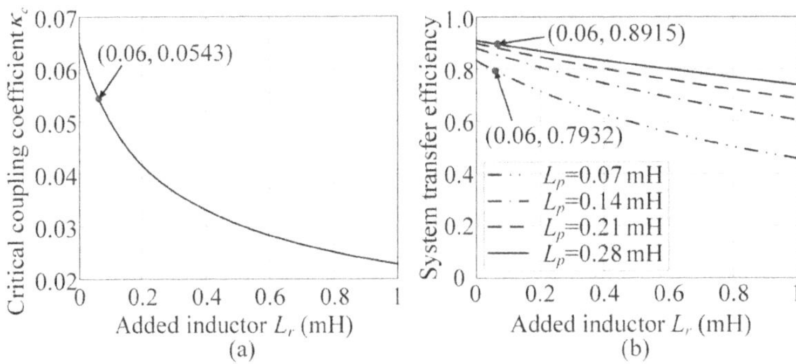

Figure 6.6 (a) κ_c against L_r; (b) η under different L_r and L_p.

The parametric design process flowchart is shown in Fig. 6.7. Initially, system resonance frequency f_0 can be selected according to the transmission range requirement and operating frequency criteria. The primary and pickup coils are configured to obtain L_p and L_s. Ferrite material can be added to the transmitting coil to strengthen

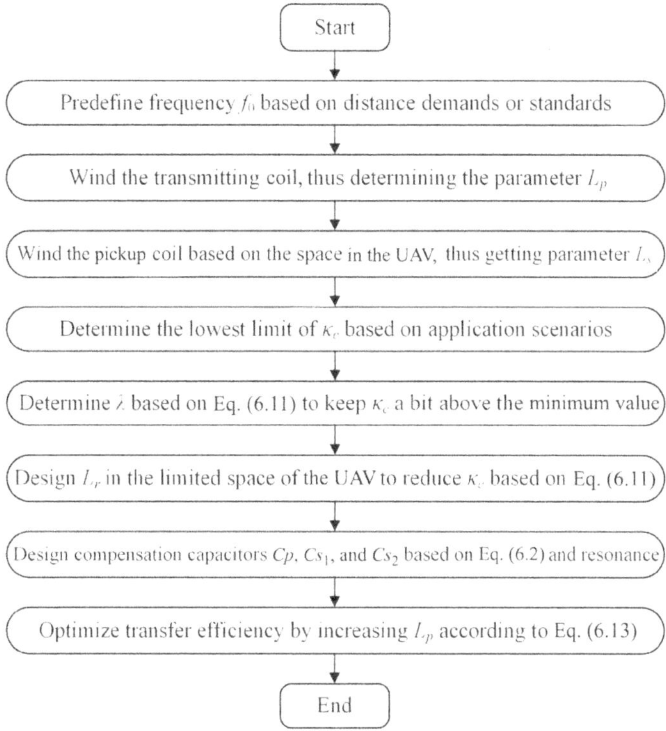

Figure 6.7 System parameter design flowchart.

the system's magnetic coupling strength. However, ferrite is not recommended for the receiving coil because the UAV side needs to be as light as possible. The minimal coupling strength of the coupling coils can be obtained from actual tests. Then the desired critical coupling κ_{c_exp} can help to make sure that this PT-WPT system enters the required range. The value of λ is chosen to keep κ_c slightly greater than this desired value κ_{c_exp}, according to Eq. (6.15). In addition, considering the limited size of the UAV, the added inductors L_r are wound to lower κ_c so that it is lower than the desired value. Based on Eq. (6.4) and the system resonance state, capacitors C_p, C_{s1}, and C_{s1} are calculated and accurately designed. Based on Eq. (6.17), the value of the primary-side coil L_p can increase the transmission efficiency without sacrificing the critical coupling coefficient. Thus, if desired, the value of L_p can be incremented to optimize the system transmission efficiency.

As an example: Example 1 represents a conventional S/S topology with the following parameters: $L_p = 73.5$ μH, $Ls = 140$ μH, $f_p = fs = 100$ kHz, $R_p = 0.39$ Ω, $R_s = 0.24$ Ω, and $R_L = 15$ Ω. In this case, the critical coupling coefficient of the system in Example 1 is 0.1724. In order to decrease the value of κ_c to be less than 0.06, the parameter that has a large influence on κ_c (capacitance ratio λ) was first chosen to be 0.6. Then, the κ_c value was decreased to 0.0638. As could be seen in Fig. 6.6(a), κ_c will reduce quicker if L_r becomes smaller than 200 μH. If L_r

becomes greater than 200 μH, the effect of L_r on κ_c is smaller. Therefore, in order to maximize this effect of L_r on the reduction in κ_c, the value of L_r should be designed to be less than 200 μH. Here, L_r is designed to be 60.5 μH. A second example, Example 2, represents the proposed S/SLDC topology with the same parameters as Example 1, with the addition of λ of 0.6 and L_r of 60.5 μH. In Example 2, κ_c has been reduced, as desired, to 0.0543. Despite the fact that this proposed system greatly extends this PT symmetry range by decreasing κ_c (from 0.1724 to 0.0543), the efficiency in this case reaches just 79.32%. It is worth noting that, as shown in Fig. 6.6(b), a bigger self-inductance L_p of the transmitting coil will increase efficiency, while κ_c can stay constant according to Eq. (6.15). Therefore, in order to maintain high system efficiency, the other parameters of Example 3 are kept consistent with Example 2, and only the L_p value is increased, to 280 μH. The efficiency of Example 3 rises to 89.15% to satisfy the wireless charging design demands. The system output properties (load power, transmission efficiency, and work frequency) of Examples 1–3 are compared later. To conclude, the proposed PT-WPT system with S/SLDC high-order topology can significantly extend the robust charging range and maintain high transmission efficiency.

The additional λ and L_r at this S/SLDC compensation network can be carefully and flexibly designed to satisfy this robust charging area demand for UAV hover charging in different scenarios. In particular, the sum of two compensation capacitors C_{s1} and C_{s2} on the secondary side is identical to the receiver-side C_s in the conventional S/S-type topology, and thus the volume and weight of the compensation capacitors are close to each other. In the proposed system, the capacitance ratio λ is the ratio of C_{s1} to the compensation capacitor C_s, which provides a convenient parameter design and revision. The merits of additional inductor L_r are small volume, low weight, and free positioning. This additional inductor can be flexibly placed in the limited space on the receiver side of the UAV to fully utilize it. In summary, the proposed S/SLDC topology has high flexibility while enhancing the robust charging range of the PT-WPT system.

6.2.3   Comparison with Other Topologies

Other than the proposed S/SLDC compensation network and this conventional S/S topology, other high-order topologies can be designed for WPT systems under the PT-symmetric energy mirroring mechanism, including the S/single inductor single capacitor (SLSC) topology and the S/double capacitor (DC) topologies [30]. Figure 6.8 shows the conventional S/S, S/SLSC, S/DC, and S/SLDC compensation networks. The deduction process of the other two compensation networks is similar to the S/SLDC topology (the detailed derivation will not be performed here). The main output characteristics of these four topologies are shown in Table 6.2.

Based on Table 6.2, the critical coupling coefficient equations for different topologies show that κ_c of the conventional S/S compensation network is the largest under the same parameter conditions, and the corresponding PT-WPT system has the smallest robust charging region. By increasing the additional inductance L_r or decreasing λ, κ_c of the S/SLSC and S/DC compensation networks is lower than that

Table 6.2 System performance comparison of PT-WPT system

	Critical coupling coefficient	Transfer efficiency
S/S compensation	$\dfrac{R_s + R_L}{\omega_0 L_s}$	$\dfrac{R_L}{R_p L_s/L_p + R_s + R_L}$
S/SLSC compensation	$\dfrac{R_s + R_L}{\omega_0 \sqrt{L_s(L_s + L_r)}}$	$\dfrac{R_L}{R_p(L_s + L_r)/L_p + R_s + R_L}$
S/DC compensation	$\dfrac{R_s + \lambda^2 R_L}{\omega_0 L_s}$	$\dfrac{\lambda^2 R_L}{R_p L_s/L_p + R_s + \lambda^2 R_L}$
Proposed S/SLDC compensation	$\dfrac{R_s + \lambda^2 R_L}{\omega_0 \sqrt{L_s(L_s + L_r)}}$	$\dfrac{\lambda^2 R_L}{R_p(L_s + L_r)/L_p + R_s + \lambda^2 R_L}$

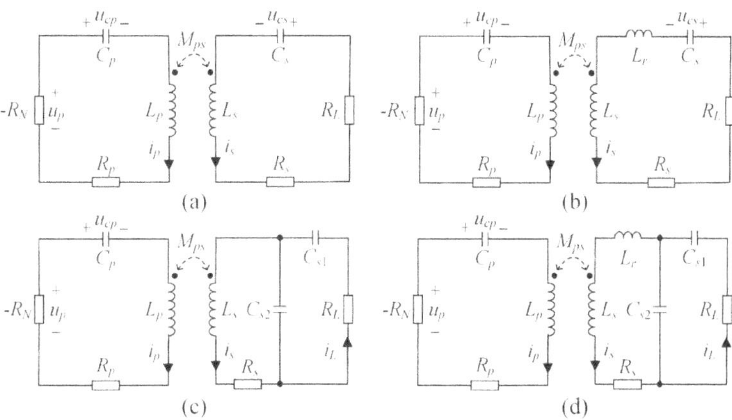

Figure 6.8 PT-WPT systems with different compensation networks (a) S/S; (b) S/SLSC; (c) S/DC; (d) The proposed S/SLDC.

of the S/S compensation network, and thus the robust charging region is enhanced to some extent. More importantly, the proposed S/SLDC compensation network integrates the benefits of the S/SLSC and S/DC networks, strengthens the system's degree of freedom through the design of additional parameters L_r and λ, and significantly reduces the value of κ_c, thus enlarging the CP region. As a result, all three high-order compensation networks extend the robust charging range of the PT-WPT system with reduced critical coupling coefficients, where the proposed S/SLDC compensation network could achieve the maximum symmetric region extension.

In order to see more clearly the impact of system parameters on output properties, Fig. 6.9(a) shows the value of κ_c and η curves for different L_r conditions within the S/SLSC topology. If this value of L_r grows, κ_c and η fall, implying that this CP region expansion reduces the transmission efficiency. Figure 6.9(b) illustrates the value of κ_c and η for various λ at this S/SPC compensation network. As the capacitance ratio λ decreases, κ_c and η decreases simultaneously.

The effect of the parameters of this S/SLDC compensation network on these output characteristics has been described previously. In summary, in comparison to the other

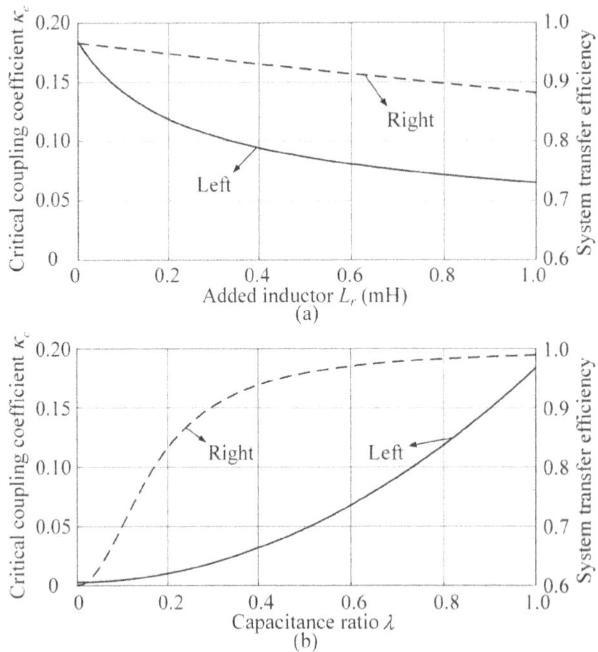

Figure 6.9 (a) κ_c and η against L_r in the S/SLSC compensation network. (b) κ_c and η against λ in the S/DC compensation network.

three types, this S/SLDC compensation network could add two more degrees of freedom and offers a wider PT symmetry region, thereby significantly extending the robust charging range. Nevertheless, the transmission efficiency is reduced when extending the system robust charging range. Therefore, suitable L_r and λ should be chosen and optimized based on various application scenarios so as to achieve a balance between the CP range and efficiency for the S/SLDC compensation network.

6.2.4 Negative Resistance Design for the PT-WPT system

In order to implement this PT-WPT system, the inverter output voltage u_p is set at the reverse phase with this current $-i_p$, and thus the power supply and the inverter are equivalently regarded as 'negative resistance'. In this work, the phase synchronization method (PSM) based on a DSP controller is employed to realize this 'negative resistance'. As illustrated in Fig. 6.10(a), this transmitting current i_p is sensed by a sensing circuit and transferred to a PWM generator to generate a pair of complementary inverter switching signals. The switching signals are passed through the driver circuit to achieve regulation of four inverter switching tubes and to generate system input voltage u_p, which is $180°$ in phase with i_p, thus realizing the PT-WPT system.

This controller programme is illustrated in Fig. 6.10(b). This transmitting-side current i_p can be sensed with the detector. The wave u'_p, at the same phase as i_p, can

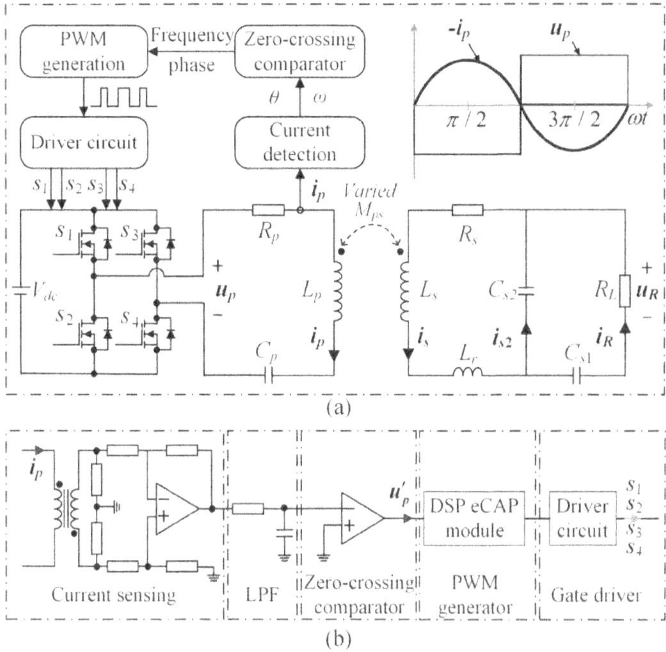

Figure 6.10 (a) System block diagram with the S/SLDC compensation network; (b) hardware implementation circuitry.

be acquired through a low-pass filter (LPF) and an over-zero comparator. The LPF is applied to reduce the interference of spurious signals in the circuit. The eCAP module of the DSP controller identifies the frequency and phase information. The drive signals produced by the controller are utilized to control the inverter switching tubes to implement the PT-WPT system. Accordingly, with the auto-frequency adjustment mechanism in this control circuit, the system can obtain robust charging behaviours in the strong coupling area. In contrast to other control methods, the entire sensing and control circuitry of the proposed system is installed at the transmitting side, which does not increment the load of the UAV, thus prolonging its flight time.

6.2.5 Experimental Verification

An experimental prototype was built to validate this practicality of this PT-WPT system with the S/SLDC topology on the basis of the system parameters shown in Table 6.3 It is depicted in Fig. 6.11. Here, Example 1 adopts the S/S compensation network for comparison to the S/SLDC compensation network.

The secondary-side SLDC compensation element consists of a receiving coil L_s, an additional resonator L_r, and compensation capacitors C_{s1} and C_{s2}. In order to compare the effect of the proposed SLDC compensation network at this pickup, the volume and weight of the receiving-side SLDC compensation element are measured. The receiver coil L_s is a planar spiral with an inner coil diameter of 5 cm and an outer coil diameter

Table 6.3 System parameters

Item	Value
Transmitter-side inductor (L_p)	73.5/285.1 μH
Transmitter-side resistor (R_p)	0.39/0.45 Ω
Transmitter-side capacitor (C_p)	34.47/8.89 nF
Receiving-side inductor (L_s)	140.7 μH
Added inductor (L_r)	60.5 μH
Series capacitor (C_{s1})	7.55 nF
Parallel capacitor (C_{s2})	5.04 nF
Receiving-side resistor (R_s)	0.34 Ω
System load (R_L)	15 Ω
Natural operating frequency (f_0)	100 kHz

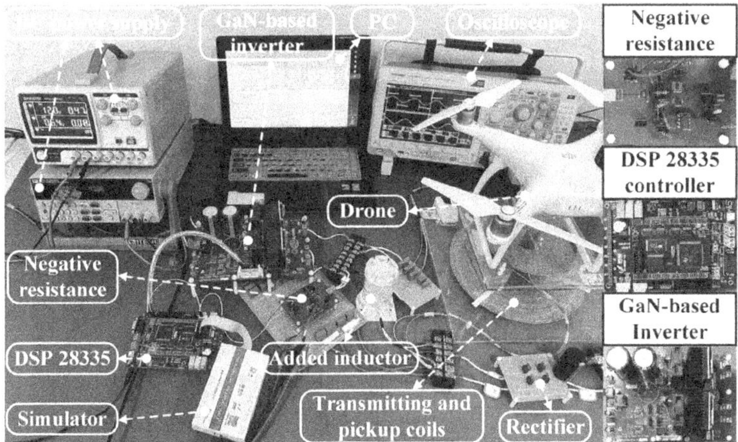

Figure 6.11 Experimental prototype.

of 15 cm. Its volume and weight are 1272.3 cm^3 and 197 g, respectively. The additional resonator L_r is selected as a solenoid-type coil with a height of 5 cm and an outer diameter of 3.5 cm. Its volume and weight are 192.4 cm^3 and 122 g, respectively. The total volume of the compensation capacitors C_{s1} and C_{s2} is 15 cm^3 and the total weight is 67 g. The added weight and volume of the SLDC compensation element on the secondary side are small and acceptable for a UAV wireless charging system. Therefore, compared with the scheme that adds pickup-side inductance to expand this CP region, the proposed topology has a big advantage. Also, the maximum transmission distance could be increased by 102% in comparison to the S/S compensation network. The detection circuit and the CP control circuit of the PT-WPT system are on the primary side, eliminating the need for an additional power converter, thus ensuring a lightweight design for the UAV.

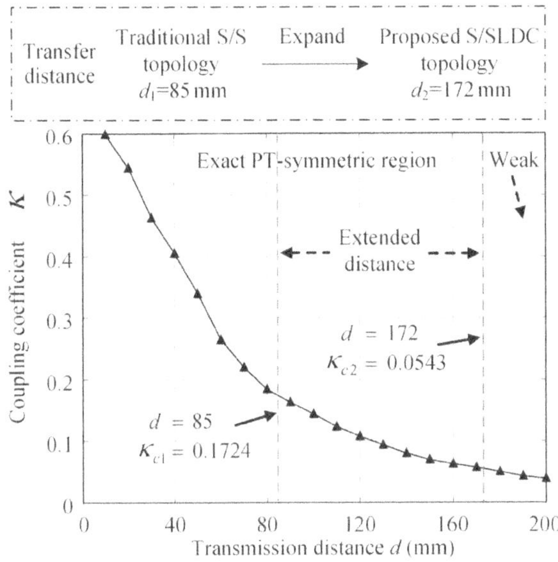

Figure 6.12 Coupling coefficient test results.

Figure 6.12 depicts the experimentally tested coupling coefficients of various air gaps. The critical coupling coefficient κ_c is 0.1724 for the conventional S/S compensation network, and the maximum transmission distance is 85 mm. By adopting this S/SLDC compensation network, the transmission distance in the robust charging area has been extended to 172 mm, accompanied by κ_c of 0.0543. It can be clearly observed that by applying this S/SLDC compensation network, this PT-WPT system's CP region (i.e. the exact PT symmetry region) is greatly extended, which is conducive to stable energy replenishment under the wide-range fluctuation of the hovering UAV.

Figure 6.13 depicts the experiment results of the PT-WPT system (Example 1) in the conventional S/S topology at four different distances. It can be seen that the system input voltage is 180 degrees out of phase with the current. The output current is stabilized at about 2.3 A at transmission distances $d = 60$ mm and $d = 80$ mm, when the system enters into this robust charging range. System load power and transmission efficiency are maintained at 79 W and 91.8%, respectively. However, if the transmission distance is increased to 100 mm, the system load current cannot be maintained robustly and the system output power changes and enters the broken state. At the same time, the system operating frequency stabilizes at 100 kHz. The experimental waveforms show the limitations of the PT-WPT conventional topology. This system urgently needs to be expanded in this robust power range to meet the application requirements of UAV hover charging.

Figure 6.14 depicts the experiment results of the PT-WPT system with the S/SLDC high-order topology at different transmission distances (90, 120, and 160 mm). It can be observed that system load current is maintained robustly regardless of the presence/absence of the rectifier bridge, and thus a stable output power is obtained. Here, load

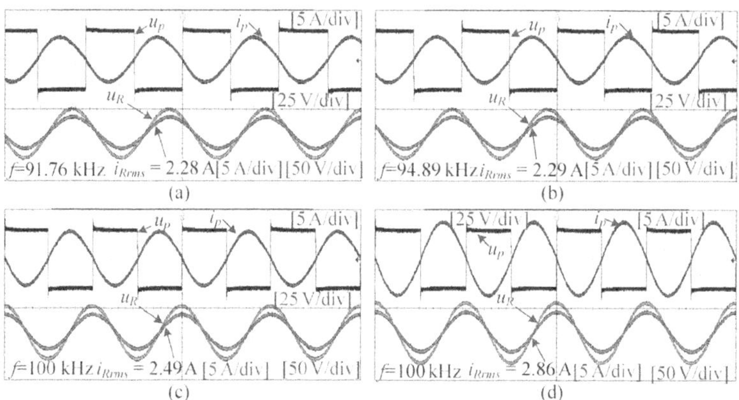

Figure 6.13 Test waveforms of S/S compensation network: (a) $d = 60$ mm; (b) $d = 80$ mm; (c) $d = 100$ mm; (d) $d = 120$ mm.

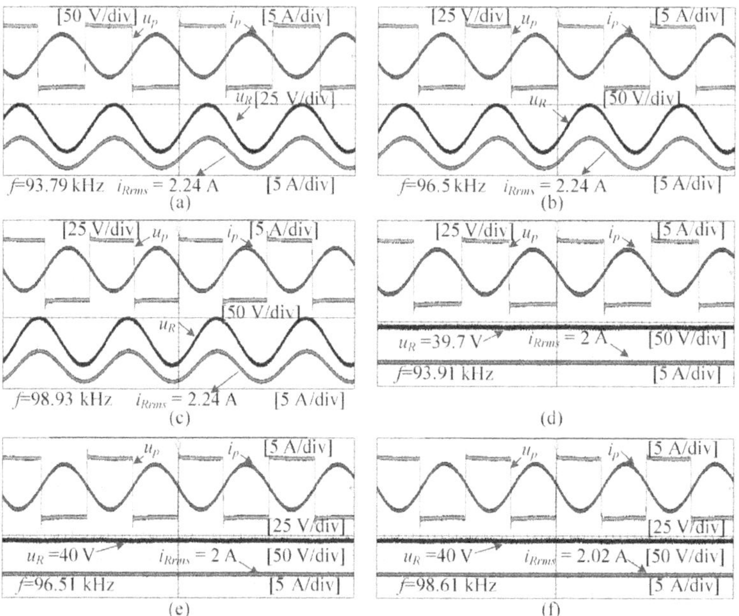

Figure 6.14 Test waveforms of the S/SLDC compensation network: (a) $d = 90$ mm; (b) $d = 120$ mm; (c) $d = 160$ mm; (d) $d = 90$ mm with full-bridge uncontrolled rectifier; (e) $d = 120$ mm with full-bridge uncontrolled rectifier; (f) $d = 160$ mm with full-bridge uncontrolled rectifier.

charging power and efficiency can be maintained at 75.3 W and 87.6%. Hence, this proposed system working under exact PT symmetry can keep system charging robust in terms of stability in UAV hover charging.

A comparison of the output properties (load power, charging efficiency, and work frequency) of the PT-WPT system in different topologies is shown in Figs. 6.15–6.17.

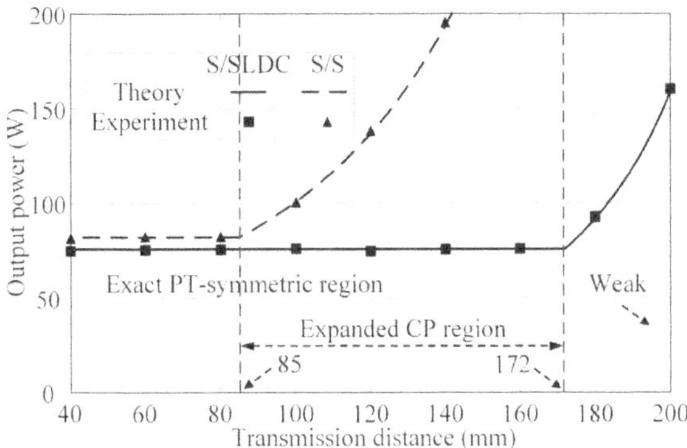

Figure 6.15 Output power at different transmission distances.

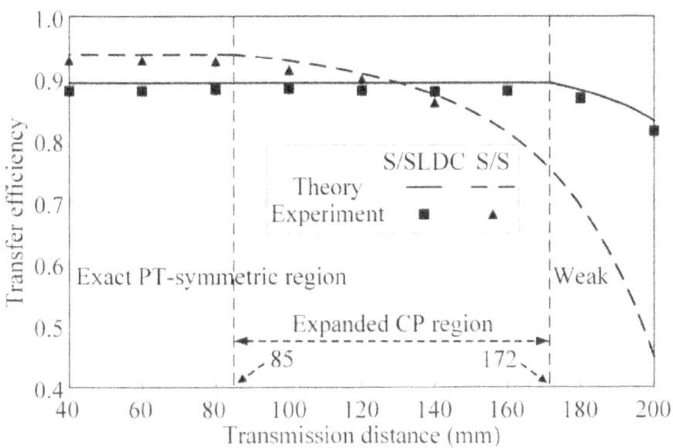

Figure 6.16 System transmission efficiency at different distances.

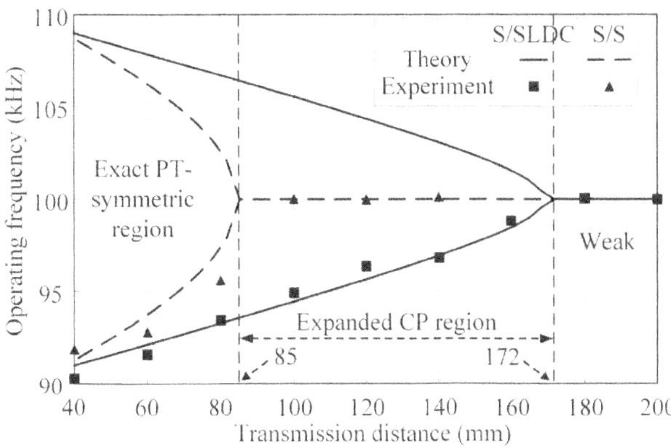

Figure 6.17 System operating frequency at different distance.

The output power in Fig. 6.15 shows that this proposed S/SLDC compensation network achieves a robust charging region expansion compared to the S/S compensation network, where a constant charging power of 75 W can be maintained even if the transmission distance varies. The results show that the transmission distance could be expanded from 85 mm to 172 mm with an extension ratio of about 102%. The transmission efficiency shown in Fig. 6.16 indicates that a stable transmission efficiency of about 87.5% can be guaranteed in the CP range. Figure 6.17 shows that under this PT-WPT system, the work frequency could be automatically adjusted to $\omega_{1,2}$. To summarize, the experiment results are in basic agreement with the theoretical calculations, and this proposed PT-WPT system with S/SLDC high-order topology can significantly expand the CP charging region and enhance the applicability of this system for applications such as drone in-flight charging.

A system power loss comparison has been carried out to compare the conventional S/S topology with the proposed higher-order S/SLDC topology. This PT-WPT system is implemented through the control of the inverter. Therefore, similar to typical inverter losses, the system losses mainly consist of MOSFET losses. The power loss of the PT-WPT system mostly consists of converter loss and loss from the coil internal resistor, as illustrated in Fig. 6.18. For this conventional S/S compensation network, system load power is 79.4 W and transmission efficiency is 91.8%; system power loss is 7.1 W. The losses from the converter and coil internal resistor are 2.6 W and 4.5 W, respectively, and the proportion of the internal resistor loss is larger at about 63.5%. The proposed S/SLDC compensation network has an output power of about 75 W with 87.5% charging efficiency and 10.7 W power loss. The internal resistor loss of 7.74 W contributes to 72.3%; converter loss is 2.96 W. There is a reduction in system charging efficiency, but it is no more than 5%. The transmission distance of the PT-WPT system with this S/SLDC compensation network is approximately more than double that of the S/S compensation network, which is a huge benefit.

The proposed PT-WPT system with S/SLDC high-order topology could be applied to UAV hover charging, which could obtain constant charging power despite the large range of coupling strength continuous perturbation. The sensing and CP control circuits of the system are located in the transmitting circuit, with no need for an additional power converter or communication module. Therefore, it ensures a

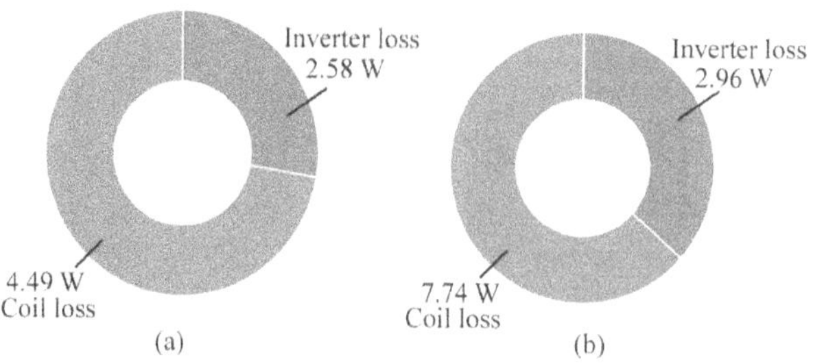

Figure 6.18 Comparison of system power losses: (a) S/S compensation network; (b) S/SLDC compensation network.

lightweight architecture for the receiver side, as well as good real-time performance of the system. The proposed high-order topology has two additional degrees of freedom that could be regulated in a flexible way to ensure greater charging efficiency while adjusting the expansion range of the CP region to satisfy the requirements of various scenarios. The additional inductor L_r provides the benefits of compact volume and low weight, and it could be flexibly located at the appropriate position on the UAV body because it does not require magnetic field coupling with the transmitting coil.

6.3 Flexible Charging Range Extension

In order to extend the symmetry range of the PT-WPT system, the S/SLDC high-order topology was proposed in the previous section to achieve extension of the symmetry region. Nevertheless, once the system parameters are fixed it will be very difficult to adjust the expanded robust charging region to suit the demands of various models of UAV and different environments. In addition, since the expansion of this PT symmetry range generally loses a certain amount of charging efficiency, it should not be extended indefinitely. A method that can adaptively adjust and expand the PT symmetry region needs to be developed to achieve a balance between the range and the efficiency required for UAV hover charging systems in different environments. Unlike other WPT implementations (e.g. EVs [31, 32], wireless motors [33, 34], consumer electronics [35, 36], and wireless lighting [37, 38]), UAV hover wireless charging faces these special challenges: how to achieve a large-range constant charging output with high flexibility for the drone battery under coupling variation and with a limited pickup-side payload.

To address these requirements, an adaptive regulation scheme for the PT symmetry region (robust charging region) with an autonomous on–off keying modulation (AOKM) strategy is introduced in this section to realize robust charging region extension and adjustment of the PT-WPT system without adding any topology units or control circuits on the receiver side [39].

The proposed AOKM strategy is theoretically analysed and these system output properties are compared under different modulation duty cycles. The design and implementation methods of the proposed scheme are described in detail, and a symmetric region adaptive regulation scheme of UAV hover charging is designed. The simulation and experimental platforms are constructed to validate the effectiveness of the proposed symmetric region flexible expansion method.

6.3.1 Autonomous On–Off Keying Modulation Scheme

Figure 6.19(a) illustrates the system circuit diagram based on the AOKM strategy. The S/S compensation network is selected for its definite features of simplicity, low weight, and high stability. These features make it particularly suited to UAV hover charging. This system has several modules: the power supply, inverter, transmitting-side circuit, receiving-side circuit, rectifier bridge, and system load. The full-bridge

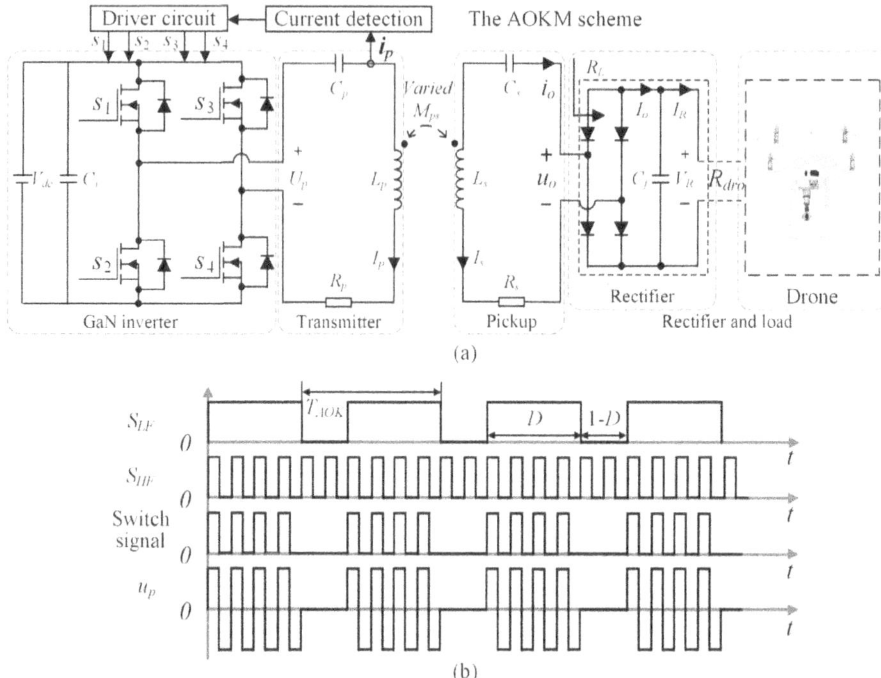

Figure 6.19 (a) Block diagram of the system with the AOKM strategy; (b) switching signals for the inverter.

inverter includes four switching tubes from S_1 to S_4, achieving DC voltage V_{dc} to AC voltage u_p. L_p, L_s, and M_{ps} represent the transmitting coil, receiving coil, and the mutual inductance. C_p, R_p and C_s, R_s represent their corresponding compensating capacitances and internal resistors, respectively. R_{eq} denotes the UAV battery equivalent load.

This PT-WPT system can obtain robust charging power for the UAV battery despite the coupling variation; the detailed derivation and analysis of output features are described in Chapter 5. It is noteworthy that the robust charging region is satisfied only when κ is higher than κ_c. Hence, κ_c is crucial for this extension and adaptive regulation of the CP region. In order to obtain a wide-range CP region, κ_c should be reduced to meet the demand for a wide range of stable charging powers for different application scenarios. The transmitting-side AOKM scheme is proposed in this section for expanding and regulating this robust charging region, which avoids the common problem of this being achieved only by a receiver-side scheme.

The proposed AOKM method can achieve the extension of the robust charging range by adjusting the primary-side duty cycle D. According to Eq. (6.22), by reducing the duty cycle D in real time, the equivalent load R_{eq} of the system will be decreased, and κ_c will also be reduced, which extends the robust charging range. At the same time, with mutual inductance perturbation, the system work frequency with the AOKM scheme will be adaptively regulated to maintain a CP output.

Figure 6.19(b) illustrates the inverter switching signals for the AOKM method. As can be seen, signals S_{HF} and S_{LF} can be employed to control the system inverter. In this case, the frequency of S_{HF} will be adaptively adjusted for the achievement of system robust charging, and S_{LF} consists of n pulses of S_{HF}, where n is normally very big. Remarkably, the inverter switching signal is logic-high only when both S_{HF} and S_{LF} are logic-high. Here, D denotes the duty cycle of the AOKM method, and T_{AOK} is the control cycle. The inverter will then cycle in turn through the active and idle modes. As depicted in Fig. 6.19(a), in the active mode I_o represents the mean output current of the rectifier bridge. The UAV battery current and load voltage are denoted I_R and V_R, respectively, where $R_{dro} = V_R/I_R$. During the T_{AOK} time period, the energy of the system after passing through the rectifier bridge should be the same as the energy received by the UAV battery:

$$E_o = V_R I_o D T_{AOK} = V_R I_R T_{AOK}. \tag{6.18}$$

A diode rectifier bridge is used on the receiving side in this system. Therefore, at the receiving side, the relationship from the amplitude of fundamental element U_{o1} of system output voltage u_o meets the below condition with respect to load voltage V_R:

$$U_{o1} = 4V_R/\pi. \tag{6.19}$$

Neglecting the energy loss of the switching tubes, the system charging power P_{out} is represented:

$$P_{out} = U_{o1}^2/(2R_{eq}) = V_R I_o. \tag{6.20}$$

Subsequently, the system equivalent load R_{eq}, by applying the AOKM method, can be concluded:

$$R_{eq} = \frac{8V_R^2}{\pi^2 V_R I_o} = D\frac{8}{\pi^2}\frac{V_R}{I_R} = D\frac{8}{\pi^2}R_{dro}. \tag{6.21}$$

As can be seen, various equivalent loads R_{eq} can be flexibly adjusted and obtained by tuning the duty cycle D of the primary side online, thus achieving the extension and regulation of this system's robust charging range while preventing an increase in the weight of the receiving side. According to the CMT modelling analysis in Chapter 5, the critical coupling coefficient κ_c, system efficiency η, and charging power P_{out} using this proposed AOKM method can be presented as Eqs. (6.22)–(6.24).

$$\kappa_c = \frac{R_s + 8DR_{dro}/\pi^2}{\omega_0 L_s}, \tag{6.22}$$

$$\eta = \frac{8DR_{dro}/\pi^2}{R_p(L_s/L_p) + R_s + 8DR_{dro}/\pi^2}, \tag{6.23}$$

$$P_o = DP_{out} = \frac{64D^2 V_{dc}^2 L_p L_s R_{dro}/\pi^2}{\pi^2\left[R_p L_s + (R_s + 8DR_{dro}/\pi^2)L_p\right]^2}. \tag{6.24}$$

It is worth noting that, according to Eq. (6.22), κ_c decreases for lower values of the duty cycle D, which enlarges this system's robust charging range and increases the degree of freedom and flexibility. According to Eqs. (6.23) and (6.24), it can be observed that system charging efficiency and load power remain unrelated to mutual inductance at a wider exact PT symmetry area. Therefore, by adopting the proposed AOKM method on the primary side, the online extension and adaptive adjustment of the CP range can be achieved without increasing this receiver-side weight, making it suitable for UAV hover charging.

6.3.2   System Output Characteristics

On the basis of the above theoretical analyses, the system output properties using the proposed AOKM method can be obtained. Based on Eqs. (6.22) and (6.21), one can see that κ_c and η depend on the load R_{dro} and the duty cycle D when the system parameters are settled. The relative output characteristic profiles at various R_{dro} and D are depicted in Fig. 6.20. It can be seen that κ_c can be decreased largely as the battery load R_{dro} decreases, thereby significantly extending the system's robust charging range. Although a lower R_{dro} would extend the CP region, the system transmission efficiency would also decrease simultaneously. Therefore, in order to achieve a trade-off between the two indices of critical coupling coefficient and transmission

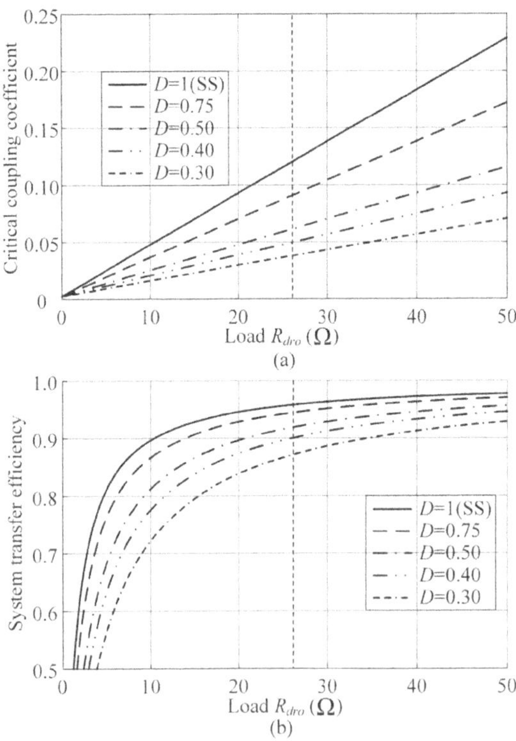

Figure 6.20 (a) κ_c and (b) η against load R_L with different duty cycles D.

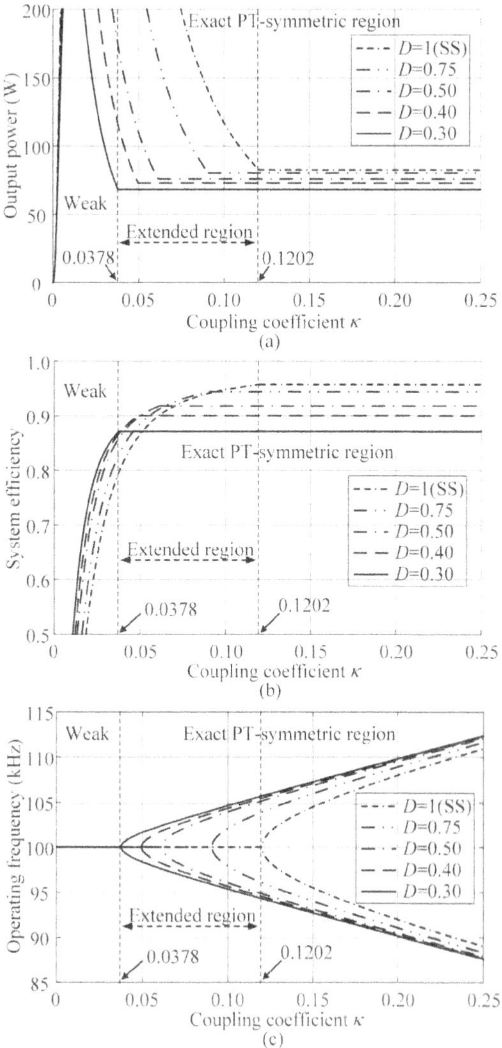

Figure 6.21 Transfer features with different D: (a) load power versus κ; (b) system efficiency versus κ; (c) working frequency versus κ.

efficiency, a suitable load value should be chosen. Here, a load value of 26 Ω is chosen to further analyse these system transfer properties.

Figure 6.21 illustrates the output properties for various duty cycles D (0.3, 0.4, 0.5, 0.75, and 1) to emphasize the adaptive expansion abilities of the proposed AOKM method for the CP area. It is shown that κ_c decreases from 0.1202 in the conventional S/S topology to 0.0378 in the proposed system with a duty cycle of 0.3, which significantly extends the robust system charging range. The figure shows that although different D leads to slight changes in charging power and efficiency, these factors remain robust against a changing charging gap. At the same time, from Fig. 6.21(c), this operating frequency will be adjusted automatically to stabilize at $\omega_{1,2}$. A lower value of D in the AOKM method can enlarge this charging gap, while it will decrease system efficiency. Therefore, a suitable D value ought to be chosen and adaptively adjusted to satisfy the needs of all kinds of UAVs and diverse scenarios over a wide range of CP areas, while maintaining high transmission efficiency as far as possible. The proposed AOKM scheme does not need complicated circuits or extra hardware, so avoiding additional receiver-side weight. The proposed AOKM scheme has tremendous convenience and flexibility for use in UAV hover charging.

6.3.3   Control Algorithm Implementation

Figure 6.22 illustrates this system block diagram with the proposed AOKM method, which enables adaptive regulation in the CP region. The system employs an S/S network topology, making it simple, lightweight, and highly reliable. The proposed AOKM method consists of two parts: adaptive frequency regulation and on–off

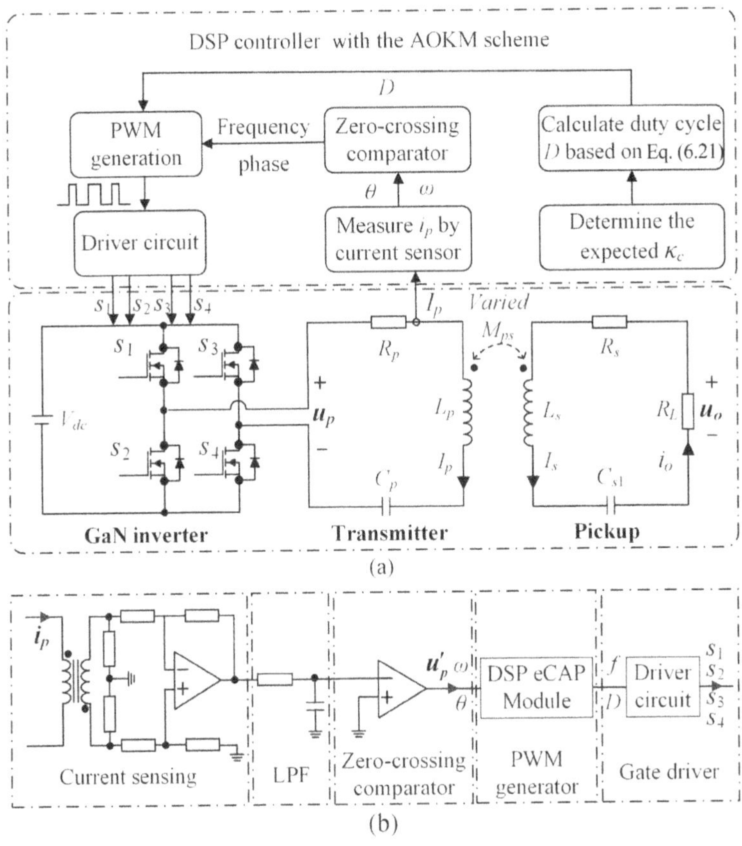

Figure 6.22 (a) System diagram with the proposed AOKM scheme. (b) Hardware implementation circuit.

keying modulation. The negative resistor can continue to use the PSM method to achieve adaptive frequency regulation. As depicted in Fig. 6.22(a), phase θ and work frequency ω of the transmitting-side current i_p of the circuit are detected and transferred to the DSP controller as it passes by this over-zero comparator. A pair of supplemental inverter switching signals are produced to ensure that the inverter output voltage u_p is 180 degrees out of phase with the transmitting-side current i_p. According to Eq. (6.24), the proposed system can acquire a CP region under mutual inductance variations in the exact PT symmetry region by adaptive frequency regulation. The duty cycle of the AOKM strategy can be tuned online once the desired κ_c is decided to adaptively enlarge this robust power range. The entire control circuit of this AOKM method is on the transmitting side, and there is no power converter or communication link on the receiver side, which is appropriate for a receiver-side lightweight design.

The signal-processing circuit of the DSP controller-based AOKM method is illustrated in Fig. 6.22(b). The transmitting-side current i_p is initially detected from the sensor, and then the wave u'_p can be produced with the LPF and over-zero comparator. This LPF contributes to reducing the influence of the ringing or spurious signal disturbances. The eCAP module of the DSP then identifies the rising edge of u'_p, containing the phase and period information of i_p. Subsequently, a high-frequency signal S_{HF} is produced in the DSP. Through the regulation of the low-frequency signal S_{LF}, comprising n S_{HF} pulses, the duty cycle of the AOKM method during the T_{AOK} cycle can be adjusted. With the determination of the duty cycle, the corresponding low-frequency signal S_{LF} is also produced. The switching signals obtained according to the S_{LF} and the S_{HF} are delivered to the gate driver of the inverter, thereby maintaining the phase difference between i_p and u_p. As a result, the working frequency of the WPT system is automatically settled in the PT symmetry region of $\omega_{1,2}$ as the mutual inductance varies. According to Eq. (6.22), κ_c would decrease as the duty cycle D is reduced. Therefore, using the proposed AOKM method, the flexible expansion and adaptive regulation of the CP range could be achieved by additional degrees of freedom.

6.3.4 Experimental Verification

To validate the practicality of this adaptive extension method for the CP region, we built a PT-WPT system prototype. The parameters are given in Table 6.4. Figure 6.23 illustrates the measured values of κ at various charging gaps with misalignments. It can be seen that the greatest transmission distance in robust charging range without the AOKM scheme is about 101 mm. As the duty cycle is adjusted, κ_c decreases, which increases the transmission distance of the robust charging range. The transmission distance could be expanded to 164 mm at $D = 0.4$, which is a 62.4% expansion.

Lateral and angular misalignment between coupling coils might occur during UAV hover charging. In order to validate the misalignment allowance capability of this AOKM method, Figs. 6.23(b) and (c) illustrate the measurable values of κ at various charging gaps and misalignments, where κ_{c1} and κ_{c4} are the critical coupling coefficients at $D = 1$ and 0.4, respectively. As can be seen, through adjusting the duty cycle

Table 6.4 System parameters

Parameters	Symbol	Value
Transmitter-side inductor	L_p	140.7 μH
Transmitter-side capacitor	C_p	18.02 nF
Transmitter-side resistor	R_p	0.24 Ω
Receiving-side inductor	L_s	285.1 μH
Series capacitor	C_s	8.89 nF
Receiving-side resistor	R_s	0.45 Ω
System load	R_L	26 Ω
Distance between coils	d	60–165 mm
System natural operating frequency	f_0	100 kHz
Period of the AOKM scheme	T_{AOK}	200/f

from 1 to 0.4, κ_c can be decreased from 0.1231 to 0.0517. This achieves improvement of the misalignment tolerance (including lateral misalignment and angular misalignment), which is consistent with the results of the theoretical analyses. Those test results validate the applicability of the presented CP range-adaptive extension method. The method is appropriate for UAV hover wireless charging under a wide range of attitude fluctuations.

Figure 6.24 illustrates the output performance (operating frequency, load charging power, and system efficiency) for various transmission distances at $D = 0.4$. According to the above theoretical analysis, this PT-WPT system will automatically stabilize at $\omega_{1,2}$ over the strong coupling range. From Fig. 6.24(a), the experimental operating frequency f of the proposed system is essentially in agreement with the theoretical value $\omega_{1,2}$. At the same time, when the transmission distance is smaller than 164 mm, a robust output power of around 73.8 W with a transmission efficiency of 88.3% will be attained no matter how the transmission distance is varied. As can be seen from Fig. 6.24, the largest transmission distance is enlarged from 101 mm without AOKM to 164 mm with $D = 0.4$ adopting the AOKM method. Significantly, the system robust charging range could be additionally regulated and enlarged by adjusting the duty cycle of the modulation strategy. Therefore, using the proposed AOKM method, the PT-WPT system could obtain a wider range of CP output, and flexible regulation of the CP region can be achieved, providing the advantages of great flexibility and low weight for drone in-flight charging applications.

To contrast its output characteristics at various duty cycles of the AOKM method, the system experiment waveforms of various transmission distances at $D = 0.4$ and 0.5 are shown in Figs. 6.25 and 6.26. Based on Eq. (6.22), this CP range can be flexibly extended by modifying the duty cycle of the AOKM method. The experimental waveforms with $D = 0.5$ at $d = 120$ mm and 150 mm are plotted in Fig. 6.25, and the waveforms with $D = 0.4$ at $d = 100$ mm and 120 mm are plotted in Fig. 6.26. As can be clearly observed, the system output current and voltage are nearly unchanged at different distances as the frequency changes. Therefore, the output

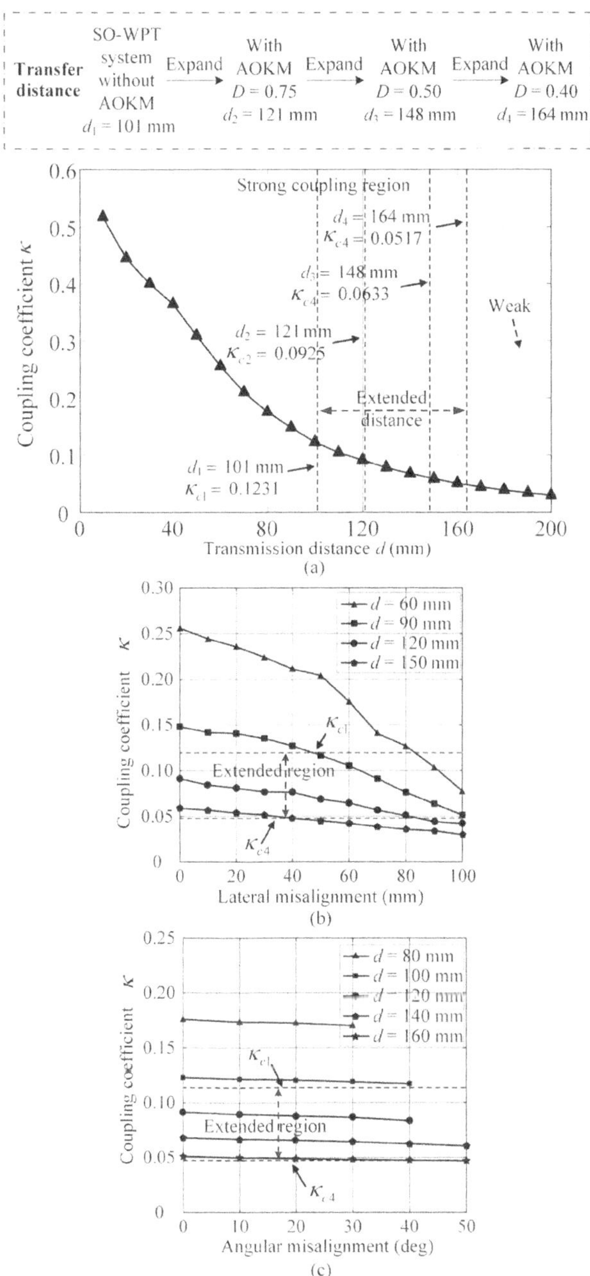

Figure 6.23 (a) Experimental values of κ under various d; (b) experimental values of κ under various lateral misalignments; (c) experimental values of κ under various angular misalignments.

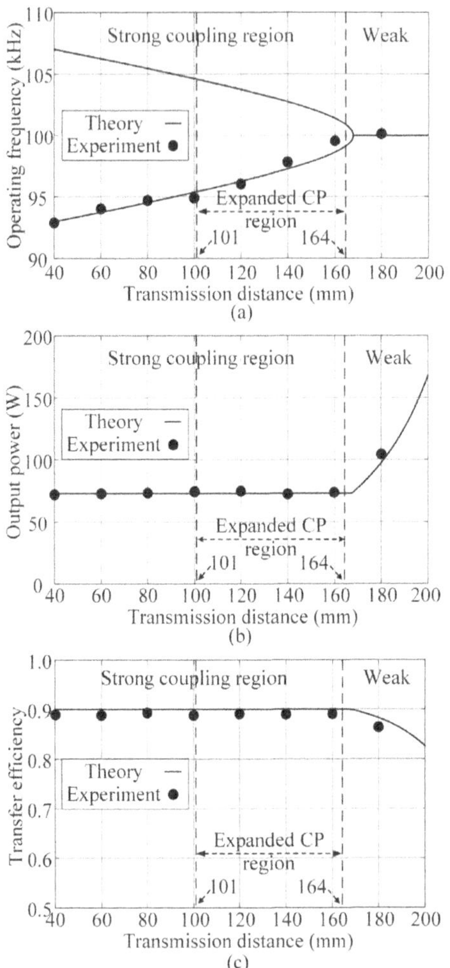

Figure 6.24 System output properties with $D = 0.4$ under different distances: (a) working frequency; (b) charging power; (c) transfer efficiency.

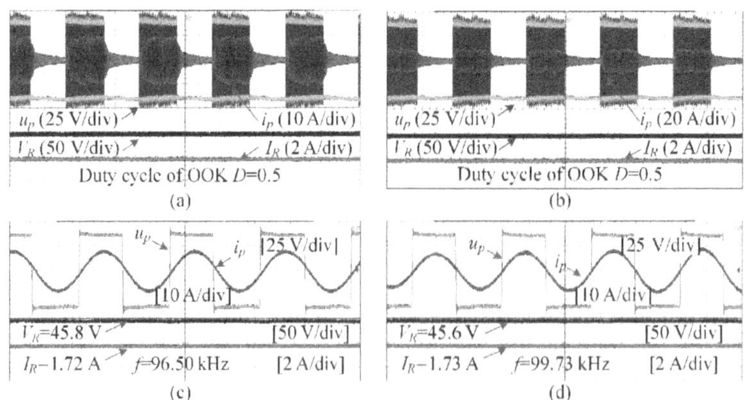

Figure 6.25 System test waveforms with $D = 0.5$: (a) $d = 120$ mm; (b) $d = 150$ mm; (c) $d = 120$ mm after zooming in; (d) $d = 150$ mm after zooming in.

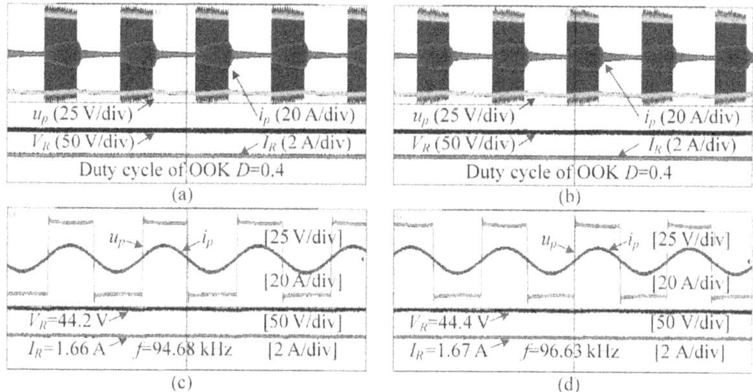

Figure 6.26 System test waveforms with $D = 0.4$: (a) $d = 100$ mm; (b) $d = 120$ mm; (c) $d = 100$ mm after zooming in; (d) $d = 120$ mm after zooming in.

power of the exact PT range is constant against transmission distance. The experimental results confirm that the proposed AOKM can extend the robust charging range with improved flexibility and robustness. The proposed CP region expansion does not require complicated network topology or additional hardware. The sensing and controlling circuitry are located on the transmitting side, eliminating the need for power converters and communication links, resulting in more speed and ensuring a lightweight UAV-side design. The proposed AOKM method for the PT-WPT system has significant benefits for UAV hover wireless charging.

References

[1] Z. Zhang, H. Pang, A. Georgiadis, and C. Cecati, 'Wireless power transfer: an overview', *IEEE Trans. Ind. Electron.*, vol. 66, no. 2, pp. 1044–1058, 2019.

[2] W. Han, K. T. Chau, C. Jiang, W. Liu, and W. H. Lam, 'Design and analysis of quasi-omnidirectional dynamic wireless power transfer for fly-and-charge', *IEEE Trans. Magn.*, vol. 55, no. 7, pp. 1–9, Jul. 2019.

[3] Y. Gu, J. Wang, Z. Liang, and Z. Zhang, 'Mutual-inductance-dynamic-predicted constant current control of LCC-P compensation network for drone wireless in-flight charging', *IEEE Trans. Ind. Electron.*, vol. 69, no. 12, pp. 12710–12719, Dec. 2022.

[4] K. Chen and Z. Zhang, 'In-flight wireless charging: a promising application-oriented charging technique for drones', *IEEE Ind. Electron. Mag.*, 2023, doi: 10.1109/MIE.2023.3246236.

[5] Z. Zhang, S. Shen, Z. Liang, S. H. K. Eder, and R. Kennel, 'Dynamic-balancing robust current control for wireless drone-in-flight charging', *IEEE Trans. Power Electron.*, vol. 37, no. 3, pp. 3626–3635, Mar. 2022.

[6] Y. Gu, J. Wang, Z. Liang, and Z. Zhang, 'Communication-free power control algorithm for drone wireless in-flight charging under dual-disturbance of mutual inductance and load', *IEEE Trans. Ind. Inform.*, vol. 20, no. 3, pp. 3703–3714, Mar. 2024.

[7] Z. Yuan, M. Saeedifard, C. Cai, Q. Yang, P. Zhang, and H. Lin, 'A misalignment tolerant design for a dual-coupled LCC-S-compensated WPT system with load-independent CC output', *IEEE Trans. Power Electron.*, vol. 37, no. 6, pp. 7480–7492, Jun. 2022.

[8] A. Hossain, P. Darvish, S. Mekhilef, K. S. Tey, and C. W. Tong, 'A new coil structure of dual transmitters and dual receivers with integrated decoupling coils for increasing power transfer & misalignment tolerance of wireless EV charging system', *IEEE Trans. Ind. Electron.*, vol. 69, no. 8, pp. 7869–7878, Aug. 2022.

[9] K. Chen and Z. Zhang, 'Rotating-coordinate-based mutual inductance estimation for drone in-flight wireless charging systems', *IEEE Trans. Power Electron.*, vol. 38, no. 9, pp. 11685–11693, Sept. 2023.

[10] Y. Gong, Z. Zhang, and S. Chang, 'Single-transmitter-controlled multiple-channel constant current outputs for in-flight wireless charging of drones', *IEEE Trans. Ind. Electron.*, vol. 71, no. 4, pp. 3606–3616, Apr. 2024.

[11] Y. Lim, H. Tang, S. Lim, and J. Park, 'An adaptive impedance-matching network based on a novel capacitor matrix for wireless power transfer', *IEEE Trans. Power Electron.*, vol. 29, no. 8, pp. 4403–4413, Aug. 2014.

[12] Y. Shao, H. Zhang, M. Liu, and C. Ma, 'Explicit design of impedance matching networks for robust MHz WPT systems with different features', *IEEE Trans. Power Electron.*, vol. 37, no. 9, pp. 11382–11393, Sept. 2022.

[13] S. Assawaworrarit, X. Yu, and S. Fan, 'Robust wireless power transfer using a nonlinear parity–time-symmetric circuit', *Nature*, vol. 546, no. 7658, pp. 387–390, Jun. 2017.

[14] Z. Hua, K. T. Chau, W. Liu, and X. Tian, 'Pulse frequency modulation for parity–time-symmetric wireless power transfer system', *IEEE Trans. Magn.*, vol. 58, no. 8, pp. 1–5, Aug. 2022.

[15] W. X. Zhong, C. Zhang, X. Liu, and S. Y. R. Hui, 'A methodology for making a three-coil wireless power transfer system more energy efficient than a two-coil counterpart for extended transfer distance', *IEEE Trans. Power Electron.*, vol. 30, no. 2, pp. 933–942, Feb. 2015.

[16] X. Shu, B. Zhang, Z. Wei, C. Rong, and S. Sun, 'Extended-distance wireless power transfer system with constant output power and transfer efficiency based on parity-time-symmetric principle', *IEEE Trans. Power Electron.*, vol. 36, no. 8, pp. 8861–8871, Aug. 2021.

[17] M. Wang, J. Feng, Y. Shi, and M. Shen, 'Demagnetization weakening and magnetic field concentration with ferrite core characterization for efficient wireless power transfer', *IEEE Trans. Ind. Electron.*, vol. 66, no. 3, pp. 1842–1851, Mar. 2019.

[18] W. Dong, C. Li, H. Zhang, and L. Ding, 'Wireless power transfer based on current non-linear PT-symmetry principle', *IET Power Electron.*, vol. 12, no. 7, pp. 1783–1791, Jun. 2019.

[19] Z. Wei and B. Zhang, 'Transmission range extension of PT-symmetry-based wireless power transfer system', *IEEE Trans. Power Electron.*, vol. 36, no. 10, pp. 11135–11147, Oct. 2021.

[20] Z. Huang, C. -S. Lam, P. -I. Mak, R. P. d. S. Martins, S. -C. Wong, and C. K. Tse, 'A single-stage inductive-power-transfer converter for constant-power and maximum-efficiency battery charging', *IEEE Trans. Power Electron.*, vol. 35, no. 9, pp. 8973–8984, Sept. 2020.

[21] S. Liu, X. Zhao, Y. Wu, L. Zhou, Y. Li, and R. Mai, 'Efficiency improvement of dual-receiver WPT systems based on partial power processing control', *IEEE Trans. Power Electron.*, vol. 37, no. 6, pp. 7456–7469, Jun. 2022.

[22] Z. Li, H. Liu, Y. Huo, J. He, Y. Tian, and J. Liu, 'High-misalignment tolerance wireless charging system for constant power output using dual transmission channels with magnetic flux controlled inductors', *IEEE Trans. Power Electron.*, vol. 37, no. 11, pp. 13930–13945, Nov. 2022.

[23] F. Xu, S.-C. Wong, and C. K. Tse, 'Overall loss compensation and optimization control in single-stage inductive power transfer converter delivering constant power', *IEEE Trans. Power Electron.*, vol. 37, no. 1, pp. 1146–1158, Jan. 2022.

[24] Y. Yang, S. Tan, and S. Y. R. Hui, 'Front-end parameter monitoring method based on two-layer adaptive differential evolution for SS-compensated wireless power transfer systems', *IEEE Trans. Ind. Inform.*, vol. 15, no. 11, pp. 6101–6113, 2019.

[25] J. Liu, G. Wang, G. Xu, J. Peng, and H. Jiang, 'A parameter identification approach with primary-side measurement for DC–DC wireless-power-transfer converters with different resonant tank topologies', *IEEE Trans. Transp. Electrif.*, vol. 7, no. 3, pp. 1219–1235, Sept. 2021.

[26] Y. Gu, J. Chen, S. Chang, and Z. Zhang, 'Constant power control against M/R with expanded PT-symmetric range for wireless in-flight charging of drones', *IEEE Trans. Magn.*, 2023, doi: 10.1109/TMAG.2023.3284826.

[27] Y. Gu, J. Wang, Z. Liang, and Z. Zhang, 'A wireless in-flight charging range extended PT-WPT system using S/single-inductor-double-capacitor compensation network for drones', *IEEE Trans. Power Electron.*, vol. 38, no. 10, pp. 11847–11858, Oct. 2023.

[28] H. A. Haus, *Waves and Fields in Optoelectronics*. Prentice-Hall, 1984.

[29] J. A. Sanders, F. Verhulst, and J. A. Murdock, *Averaging Methods in Nonlinear Dynamical Systems*, 2nd ed. Springer, 2007.

[30] Y. Gu, Q. Zhu, L. Jia, G. Li, and Z. Zhang, 'Optimized high-order compensation topology for PT-WPT systems with expanded constant power region', *IEEE Trans. Magn.*, 2023, doi: 10.1109/TMAG.2023.3287481.

[31] P. Zhang, M. Saeedifard, O. C. Onar, Q. Yang, and C. Cai, 'A field enhancement integration design featuring misalignment tolerance for wireless EV charging using LCL topology', *IEEE Trans. Power Electron.*, vol. 36, no. 4, pp. 3852–3867, Apr. 2021.

[32] X. Dai, J. Jiang, and J. Wu, 'Charging area determining and power enhancement method for multiexcitation unit configuration of wirelessly dynamic charging EV system', *IEEE Trans. Ind. Electron.*, vol. 66, no. 5, pp. 4086–4096, May 2019.

[33] C. Jiang, K. T. Chau, C. H. T. Lee, W. Han, W. Liu, and W. H. Lam, 'A wireless servo motor drive with bidirectional motion capability', *IEEE Trans. Power Electron.*, vol. 34, no. 12, pp. 12001–12010, Dec. 2019.

[34] W. Han, K. T. Chau, Z. Hua, and H. Pang, 'An integrated wireless motor system using laminated magnetic coupler and commutative-resonant control', *IEEE Trans. Ind. Electron.*, vol. 69, no. 5, pp. 4342–4352, May 2022.

[35] Y. Gu, J. Wang, Z. Liang, Y. Wu, C. Cecati, and Z. Zhang, 'Single-transmitter multiple-pickup wireless power transfer: advantages, challenges, and corresponding technical solutions', *IEEE Ind. Electron. Mag.*, vol. 14, no. 4, pp. 123–135, Dec. 2020.

[36] C. Liu, C. Jiang, J. Song, and K. T. Chau, 'An effective sandwiched wireless power transfer system for charging implantable cardiac pacemaker', *IEEE Trans. Ind. Electron.*, vol. 66, no. 5, pp. 4108–4117, May 2019.

[37] X. Qu, W. Zhang, S.-C Wong, and C. K. Tse, 'Design of a current-source-output inductive power transfer LED lighting system', *IEEE J. Emerg. Sel. Topics Power Electron.*, vol. 3, no. 1, pp. 306–314, Mar. 2015.

[38] W. Liu, K. T. Chau, C. H. T. Lee, C. Jiang, W. Han, and W. H. Lam, 'A wireless dimmable lighting system using variable-power variable-frequency control', *IEEE Trans. Ind. Electron.*, vol. 67, no. 10, pp. 8392–8404, Oct. 2020.

[39] Y. Gu, J. Wang, Z. Liang, and Z. Zhang, 'Flexible constant-power range extension of self-oscillating system for wireless in-flight charging of drones', *IEEE Trans. Power Electron.*, 2024, doi: 10.1109/TPEL.2024.3440963.

Index

Printed by Integrated Books International,
United States of America